D0018619

RELATIVITY SIMPLY EXPLAINED

RELATIVITY SIMPLY EXPLAINED

MARTIN GARDNER

Illustrated by
Anthony Ravielli

DOVER PUBLICATIONS, INC.
Mineola, New York

For Billie, a relative

Copyright

Bibliographical Note

This Dover edition, first published in 1997, is a corrected and enlarged republication of the work originally published in 1962 by the Macmillan Company, New York, under the title *Relativity for the Million* and revised in 1976 by Vintage Books (Random House), New York, under the title *The Relativity Explosion*. The Dover edition makes new corrections, restores all the 1962 illustrations in their original color, and adds a new Introduction and a new Postscript by the author.

Library of Congress Cataloging-in-Publication Data

Gardner, Martin, 1914–
[Relativity explosion]
Relativity simply explained / Martin Gardner.
p. cm.
Originally published: The relativity explosion. Rev. updated ed. New York : Vintage Books, 1976.
Includes index.
ISBN-13: 978-0-486-29315-8
ISBN-10: 0-486-29315-7
1. Relativity (Physics) I. Title.
QC173.57.G37 1996
530.1'4—dc20 96-28034
CIP

Manufactured in the United States by Courier Corporation
29315709
www.doverpublications.com

Contents

Introduction to the Dover Edition

Although here and there in this book there are minor corrections and additions, the main addition is a postscript. I have tried in that final chapter to update the book with brief discussions of major events in relativity theory and its confirmations during the twenty years that have passed since the revised second edition was published.

I never cared for the book's two earlier titles: *Relativity for the Million* (1962), an echo of Lancelot Hogben's popular *Mathematics for the Million*, or *The Relativity Explosion* (1976). Hayward Cirker, president of Dover Publications, suggested the present title. I much prefer it because it conveys exactly what the book is intended to be.

The second edition omitted many of Tony Ravielli's illustrations, and dropped the blue color overlays from the pictures that were used. I am happy to report that his original art, with its color, is here restored.

Introduction to the 1976 Edition

This is a much revised, updated version of my book *Relativity for the Million*, published in 1962. Two entirely new chapters have been added: Chapter 7 to review the latest tests of Einstein's theory of gravity, Chapter 11 to report on three stupendous new astronomical discoveries—quasars, pulsars, and possible black holes—that are intimately connected with relativity. The last chapter has been greatly expanded to provide an obituary for the steady-state theory of the cosmos, and to place more emphasis on the currently fashionable pulsating models. John Archibald Wheeler's vision of a universe emerging from superspace, expanding, contracting, and re-entering superspace is shown to be a sophisticated elaboration of a model proposed by Edgar Allan Poe. Throughout the book there have been extensive revisions.

So many popular books on relativity had been written before 1962 that readers may wonder why I then wanted to write another one. I had three reasons:

1. The best introductions to elementary relativity had been written many years before 1962, and all of them were out of date. So many exciting new developments had taken place, all bearing on relativity theory, that I was convinced it was time for a new introductory book that would include this new material.

2. It was a challenge to try to explain once more, in a simple and entertaining way, the main aspects of Einstein's revolutionary theory. What did Einstein mean when he wrote "Newton, forgive me"? In my opinion anyone today who does not understand what he meant is as deficient in his education as someone who, a hundred years ago, knew nothing about Isaac Newton's great contributions to science. I myself was eager to learn more about relativity. Is there any better way to teach oneself a topic than to write a book about it?

3. No popular book on relativity had been illustrated so elaborately. Anthony Ravielli's brilliant graphic art alone sets this book apart from earlier introductions.

If the reader wonders why the book contains no chapter on the philosophical consequences of relativity, it is because I am firmly persuaded that in the ordinary sense of the word "philosophical," relativity *has* no consequences. For the theory of knowledge and the philosophy of science it obviously has implications, chiefly through its demonstration that the mathematical structure of space and time cannot be determined without observation and experiment. But as far as the great traditional topics of philosophy are concerned—God, immortality, free will, good and evil, and so on—relativity has absolutely nothing to say. The notion that relativity physics supports the avoidance of value judgments in anthropology, for example, or a relativism with respect to morals, is absurd. Actually, relativity introduces a whole series of new "absolutes."

It is sometimes argued that relativity theory makes it more difficult to think that outside our feeble minds there is a "huge world" possessing an orderly structure that can be described in part by scientific laws. "As the subject [relativity] developed," writes the English astronomer James Jeans in his book *The Growth of Physical Science*, "it became clear that the phenomena of nature were determined by us and our experience rather than by a mechanical universe outside us and independent of us."

Jeans's idealism is a respectable metaphysical attitude, and there are even aspects of quantum mechanics that may bear upon it, but it receives not the slightest support from relativity. I will not argue the point here. It has been forcibly made by almost every modern philosopher of science. The interested reader will find a particularly clear discussion in Chapter 7, "Metaphysical Interpretations of Relativistic Physics," in Philipp Frank's *Philosophy of Science*.

Is there not something enormously narcissistic about the notion that we humans, with our crude little brains so recently evolved from the brains of beasts, are somehow partial creators of the universe? That nothing could be more distant from Einstein's own humility you will see at once when you read the beautiful quotation that follows as this book's epigraph.

M.G.

Out yonder there was this huge world, which exists independently of us human beings and which stands before us like a great, eternal riddle, at least partially accessible to our inspection and thinking. The contemplation of this world beckoned like a liberation . . . .

Albert Einstein,
Autobiographical Notes

1

Absolute or Relative?

Two sailors, Joe and Moe, were cast away on a deserted island. Several years went by. One day Joe found a bottle that had washed ashore. It was one of those new king-size bottles of Coca-Cola. Joe turned pale.

"Hey, Moe!" he shouted. "We've shrunk!"

There is a serious lesson to be learned from this joke. The lesson is: There is no way of judging the size of an object except by comparing it with the size of something else. The Lilliputians thought Gulliver a giant. The Brobdingnagians thought Gulliver tiny. Is a billiard ball large or small? Well, it is extremely large *relative* to an atom, but extremely small *relative* to the earth.

Jules Henri Poincaré, a famous nineteenth-century French mathematician who anticipated many aspects of relativity theory, once put it in this way (scientists call his way of putting it a "thought experiment": an experiment that can be imagined but not actually performed). Suppose, he said, that during the night, while you were sound asleep, everything in the universe became a thousand times larger than before. By everything, Poincaré meant *everything:* electrons, atoms, wavelengths of light, you yourself, your bed, your house, the earth, the sun, the stars. When you awoke, would you be able to tell that anything had changed? Is there any experiment you could perform that would prove you had altered in size?

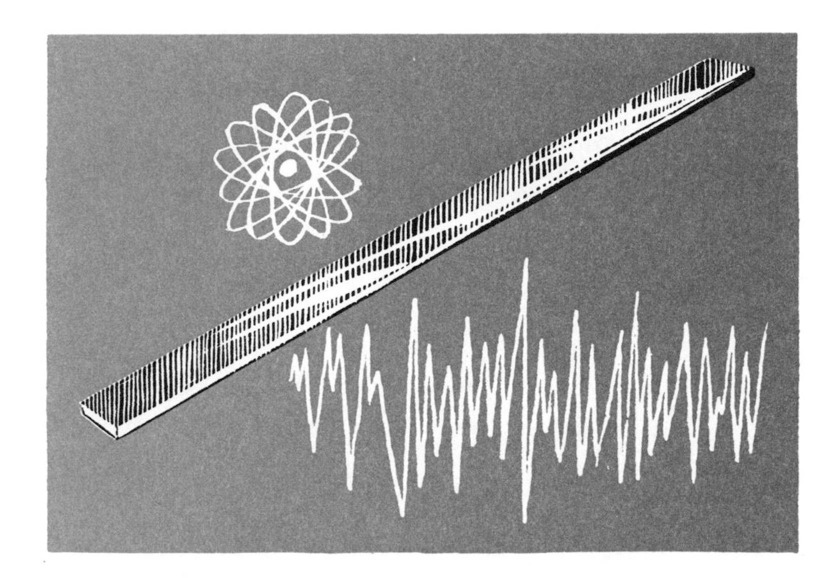

No, said Poincaré, there is no such experiment. In fact, the universe really would be the same as before. It would be meaningless even to say it had grown larger. "Larger" means larger in relation to something else. In this case there is no "something else." It would be just as meaningless, of course, to say that the entire universe had shrunk in size.

Size, then, is relative. There is no *absolute* way to measure an object and say that it is absolutely such-and-such a size. It can be measured only by applying other sizes, such as the length of a yardstick or meter rod. But how long is a meter rod? Originally it was defined as one ten-millionth of the distance from the earth's equator to one of its poles. This soon gave way to the length of a platinum bar kept in a cellar near Paris. Today it is defined as the distance light travels through a vacuum in one 299,792,458th of a second. How is a second defined? It is 9,192,631,770 vibrations of a cessium atom excited by microwaves. Of course, if everything in the universe were to grow larger or smaller in the same proportion, including the distance light travels in a second, there would still be no experimental way to detect the change.

The same is true of periods of time. Does it take a "long" or "short" time for the earth to make one trip around the sun? To a small child, the time from one Christmas to the next seems like an eternity. To a geologist, accustomed to thinking in terms of millions of years, one year is but a fleeting instant. A period of time, like distance in space, is impossible to measure without comparing it to some other period of time. A year is measured by the earth's

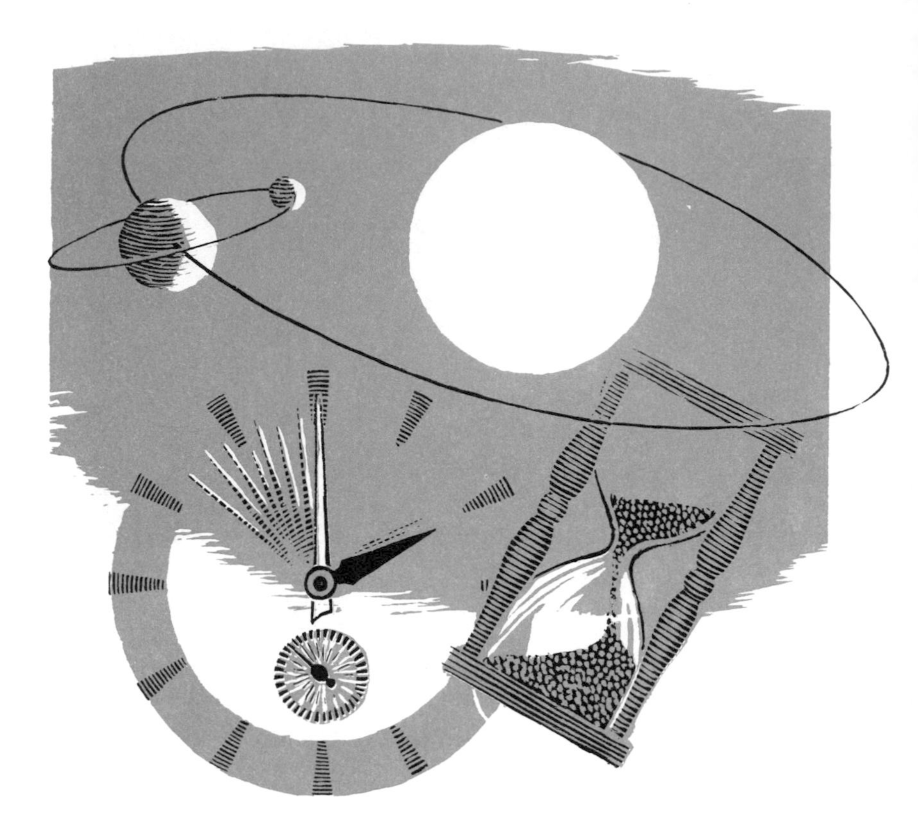

period of revolution around the sun; a day by the time it takes the earth to rotate once on its axis; an hour by the time it takes the long hand of a clock to make one revolution. Always one period of time is measured by comparing it with another.

There is a famous science-fiction story by H. G. Wells called "The New Accelerator." It teaches the same lesson as the joke about the two sailors, only the lesson is about time instead of space. A scientist discovers a way to speed up all the processes of his body. His heart beats more rapidly, his brain operates faster, and so on. You can guess what happens. The world seems to slow down to a standstill. The scientist walks outside, moving slowly so the friction of the air will not set fire to his pants. The street is filled with human statues. A man is frozen in the act of winking at two passing girls. In the park, a band plays with a low-pitched, wheezy rattle. A bee buzzes through the air with the pace of a snail.

Let us try another thought experiment. Suppose that at a certain instant everything in the cosmos begins to move at a slower speed, or a faster speed,* or perhaps stops entirely for a few million years, then starts up again. Would the change be perceivable? No, there is no experiment by which it could be detected. In fact, to say that such a change had occurred would be meaningless. Time, like distance in space, is relative.

* We disregard in this book, as in most popular science writing, a technical distinction between speed and velocity. Speed is a *scalar* (a magnitude expressed by one variable), whereas velocity is a *vector*, calling for a specified direction as well as distance divided by time. If a car spirals around a hill on its way to the top, its speed at any moment is indicated by the speedometer. But its average velocity from the bottom to the top of the hill is much less than its average speed because it is obtained by dividing the distance along the vector line (a straight line from bottom to top) by the time.

Many other concepts familiar in everyday life are relative. Consider "up" and "down." In past ages it was hard for people to understand why a man on the opposite side of the earth was not hanging upside down, with all the blood rushing to his head. Children today have the same difficulty when they first learn that the earth is round. If the earth were made of transparent glass and you could look straight through it with a telescope, you would in fact see people standing upside down, their feet sticking to the glass. That is, they would appear upside down *relative to you*. Of course, *you* would appear upside down relative to *them*. On the earth, "up" is the direction that is away from the center of the earth. "Down" is toward the center of the earth. In interstellar space there is no absolute up or down, because there is no planet available to serve as a "frame of reference."

Imagine a spaceship on its way through the solar system. It is shaped like a giant doughnut and is rotating so that centrifugal force creates an artificial gravity field. Inside the ship, spacemen can walk about the outer rim of the doughnut as if it were a floor. "Down" is now *away* from the center of the ship, "up" is *toward* the center: just the opposite of how it is on a rotating planet.

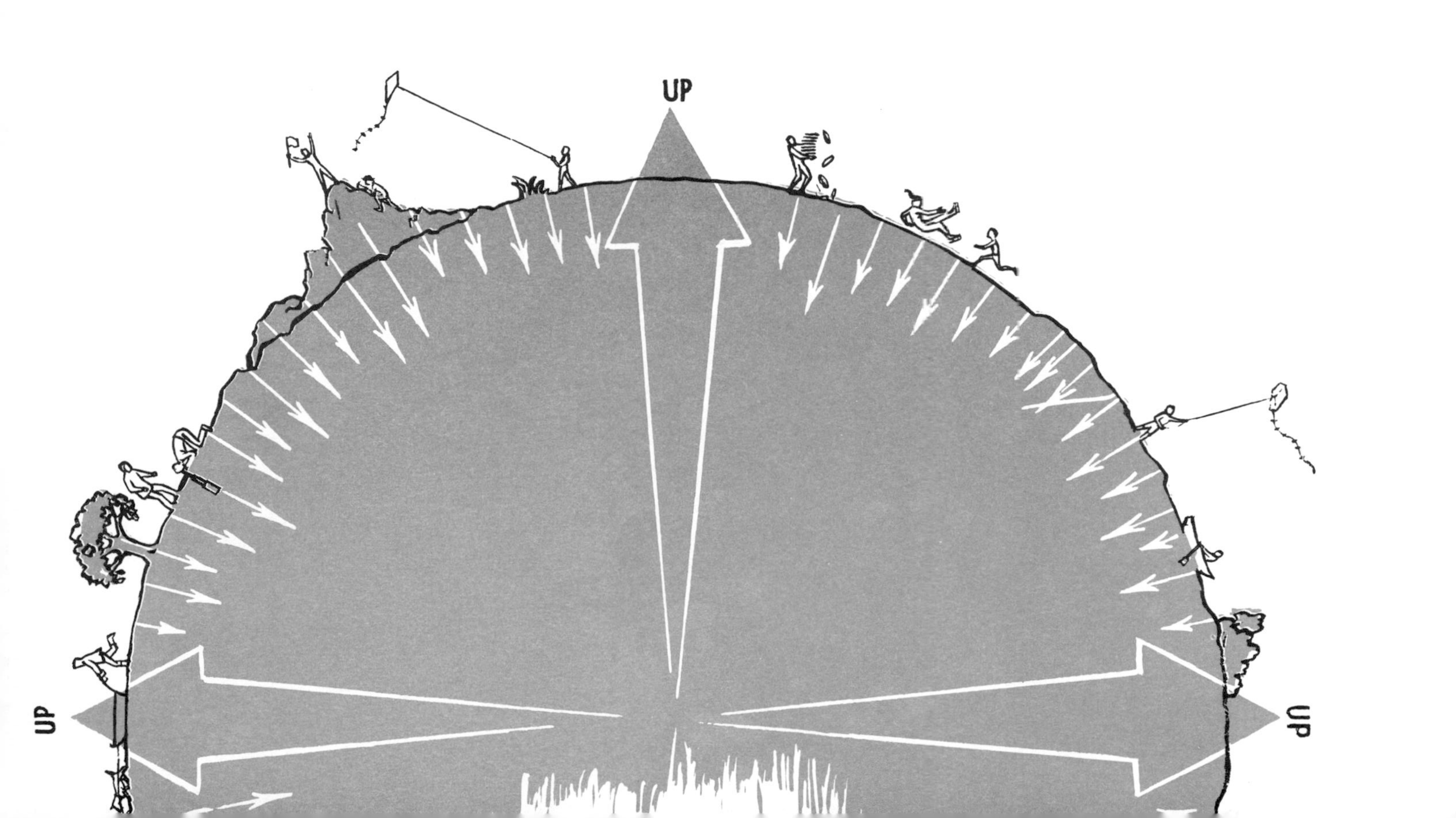
UP
UP
UP

So you see, there are no absolute "ups" and "downs" in the universe. Up and down are directions relative to the direction in which a gravitational field is acting. It would be meaningless to say that while you were asleep the entire cosmos turned upside down, because there is nothing to serve as a frame of reference for deciding which position the cosmos has taken.

Another type of change that is relative is the change of an object to its mirror image. If a capital R is printed in reverse form like this, Я, you recognize it immediately as the mirror image of an R. But if the entire universe (including you) suddenly became its mirror image, there would be no way that you could detect such a change. Of course, if only one person became his mirror image (H. G. Wells wrote a story about this also, "The Plattner Story") while the cosmos remained the same, then it would seem to him as if the cosmos had reversed. He would have to hold a book up to a mirror to read it, the way Alice behind the looking-glass managed to read the reversed printing of "Jabberwocky" by holding the poem up to a mirror. But if *everything* reversed, there would be no experiment that could detect the change. It would be just as meaningless to say that such a reversal had occurred as it would be to say that the universe had turned upside down or doubled in size.

Is motion absolute? Is there any type of experiment that will show positively whether an object is moving or standing still? Is motion another relative concept that can be measured only by comparing one object with another? Or is there something peculiar about motion, something that makes it different from the relative concepts just considered?

Stop and think carefully about this for a while before you go on to the next chapter. It was in answer to just such questions that Einstein developed his famous theory of relativity. This theory is so revolutionary, so contrary to common sense that even today there are thousands of scientists (including physicists) who have as much difficulty understanding its basic concepts as a child has in understanding why the people of China do not fall off the earth.

If you are young, you have a great advantage over these scientists. Your mind has not yet developed those deep furrows along which thoughts so often are forced to travel. But whatever your age, if you are willing to flex your mental muscles, there is no reason why you cannot learn to feel at home in the strange new world of relativity.

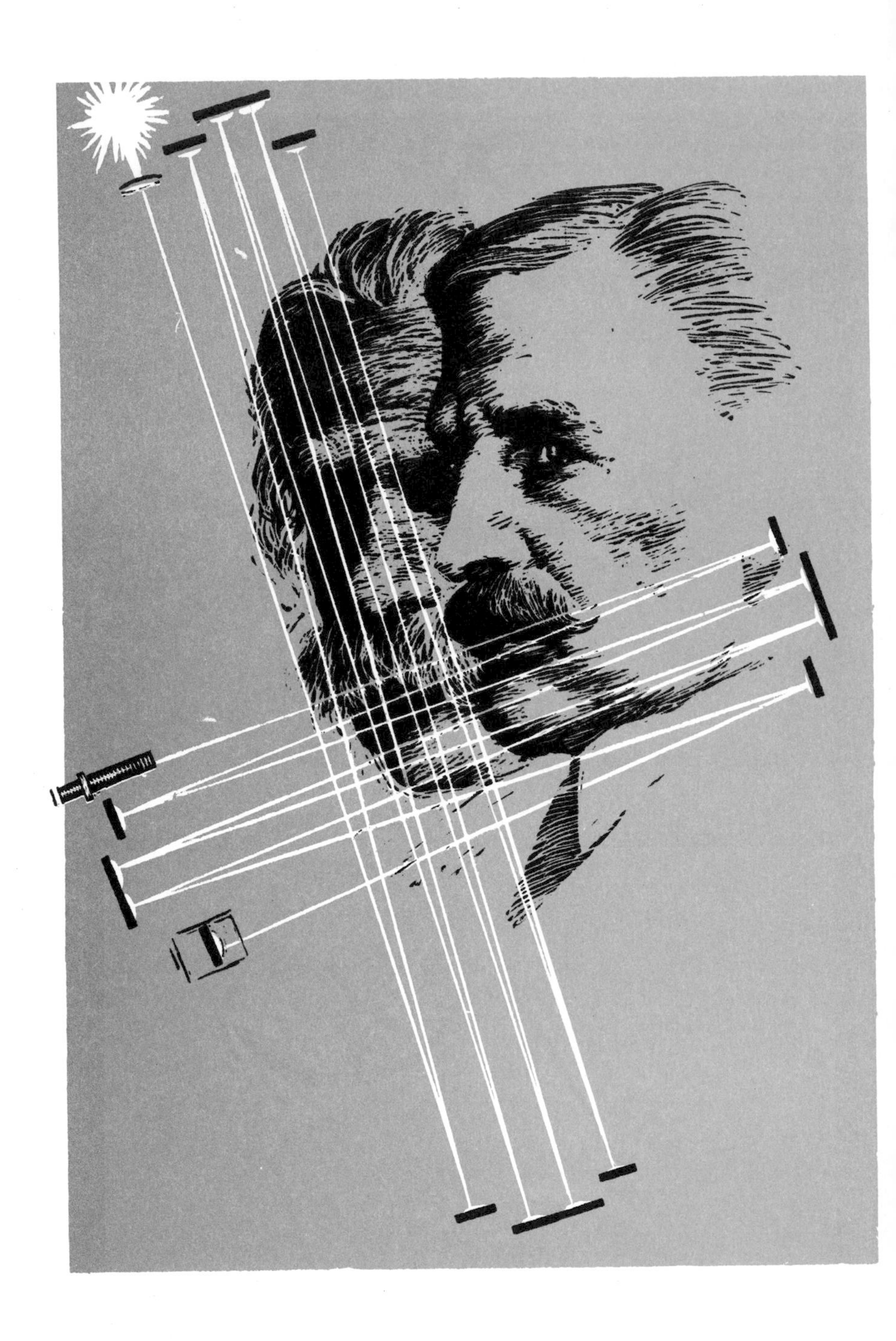

2
The Michelson-Morley Experiment

Is motion relative? After some first thoughts you may be inclined to answer, "Of course it is!" Imagine a train moving north at 100 kilometers per hour. On the train a man walks south at 4 kilometers per hour. In what direction is he moving and at what speed? It is immediately obvious that this question cannot be answered without choosing a frame of reference. Relative to the train, the man moves south at 4 kilometers per hour. Relative to the ground, he moves north at $100 - 4 = 96$ kilometers per hour.

Can we say that the man's "ground speed" (96 kilometers per hour) is his true, absolute speed? No, because there are other, larger frames of reference. The earth itself is moving. It both rotates and swings around the sun. The sun, with all its planets, speeds through the galaxy. The galaxy rotates and moves relative to other galaxies. The galaxies in turn form galactic clusters that move relative to each other. No one really knows how far this chain of motions can be carried. There is no apparent way to chart the absolute motion of anything; that is to say, there is no fixed, final frame of reference by which all motions can be measured.* Motion and rest, like large and small, slow and fast, up and down, left and right, seem to be completely relative. There is no way to measure the motion of one object except by comparing it with the motion of some other object.

Alas, it is not so simple! If this were all there is to say about the relativity of motion, there would have been no need for Einstein to develop his theory of relativity. Physicists would have had the theory all along!

The reason it is not simple is this: there appear to be two very easy ways to detect absolute motion. One method makes use of the speed of light; the other makes use of various inertial effects that occur when a moving object alters its path or velocity. Einstein's special theory of relativity deals with the first, his general theory of relativity with the second. In this and the next two chapters the first method that might serve as a clue to absolute motion, the method that makes use of the speed of light, will be considered.

In the nineteenth century, before the time of Einstein, physicists thought of space as containing a kind of fixed, invisible substance called the *ether*. Often it was called the "luminiferous ether," meaning that it was the bearer of light waves. It filled the entire universe. It penetrated all material substances. If all the air were pumped out of a glass bell jar, the jar would still be filled—filled with ether. Otherwise, how could light travel through the vacuum? Light is a wave motion; there had to be something there to transmit the waving. The ether itself, although it must vibrate, seldom (if ever) would move with respect to material objects; rather, all objects would move through it, like the movement of a sieve through water. The absolute motion of a star, planet, or any object whatever was (so these early physicists were convinced) simply its motion with respect to this motionless, invisible, etherial sea.

But, you may ask, if the ether is an invisible, nonmaterial substance—a substance that cannot be seen, heard, felt, smelled or tasted—how can the

* Since that sentence was written, a way has been found to measure the earth's quasi-absolute motion relative to the black-body radiation that permeates our universe. This will be covered in Chapter 12.

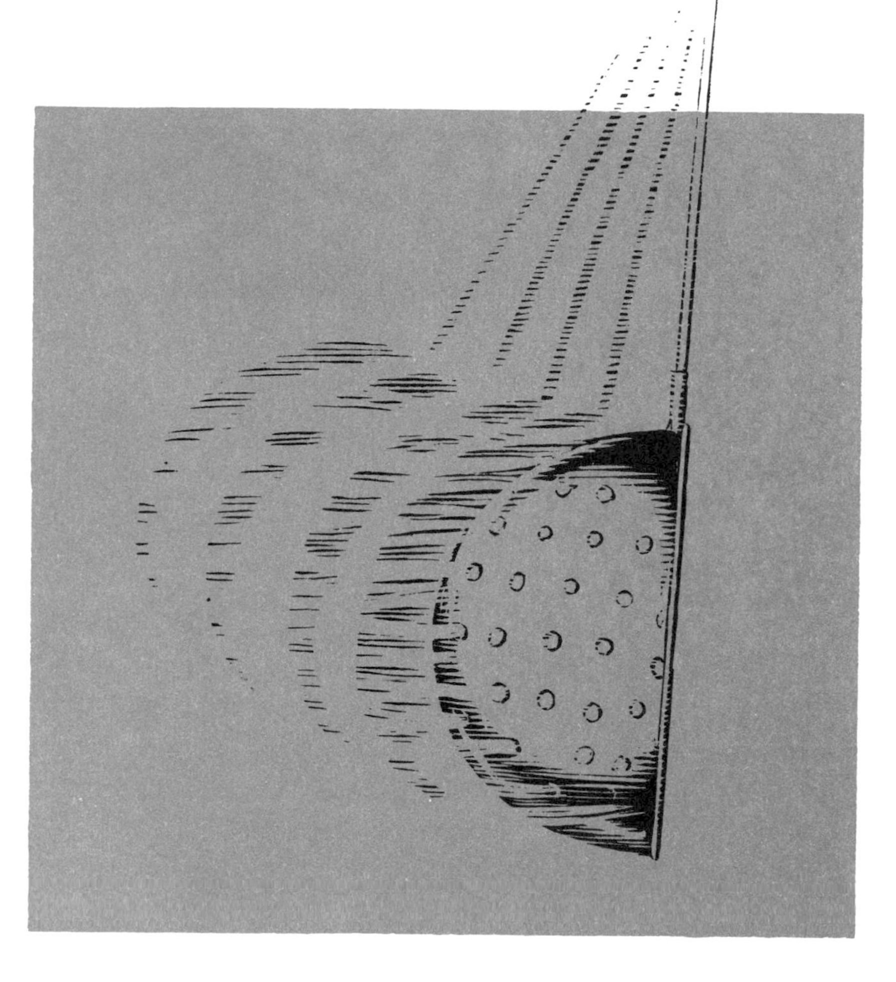

movement of, say, the earth ever be measured with respect to it? The answer is simple. The measurement can be made by comparing the earth's motion with the motion of a beam of light.

To understand this, consider for a moment the nature of light. Actually, light is only the small visible portion of a spectrum of electromagnetic radiation which includes radio waves, radar waves, infrared light, ultraviolet light, and gamma rays. Everything said about light in this book applies equally to any type of electromagnetic wave, but "light" is a shorter term than "electromagnetic wave," so this term will be used throughout. Light is a wave motion. To think of such a motion without thinking also of a material ether seemed to the early physicists as preposterous as thinking about water waves without thinking of water.

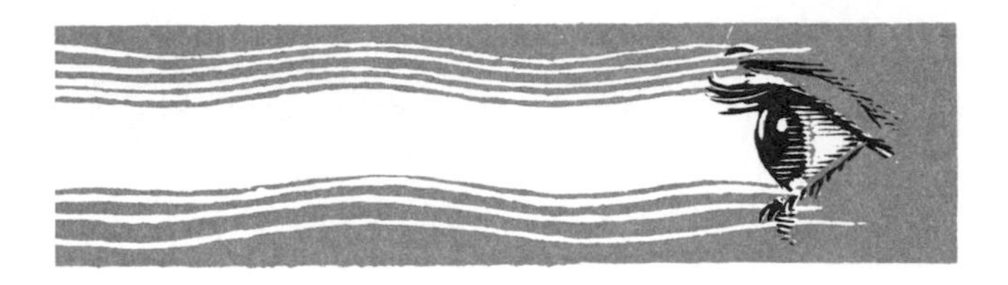

If a bullet is fired straight ahead from the front of a moving jet plane, the ground speed of the bullet is faster than if it were fired from a gun held by someone on the ground. The ground speed of the bullet fired from the plane is obtained by adding the speed of the plane to the speed of the bullet. In the case of light, however, the velocity of a beam is not affected by the speed of the object that sends out the beam. This was strongly indi-

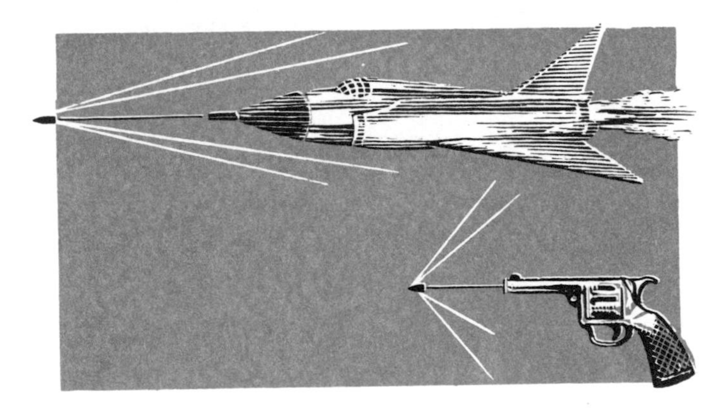

cated by experiments in the late nineteenth and early twentieth centuries, and has since been amply confirmed, especially by recent tests on the decay of neutral pi mesons. One famous test was made by Russian astronomers in 1955, using light from opposite sides of the rotating sun. One edge of our sun is always moving toward us, the other edge always moving away. It was found that light from both edges travels to the earth with the same velocity. Similar tests had been made decades earlier with light from revolving double stars. Regardless of the motion of its source, the speed of light through empty space is always the same: about 299,800 kilometers (186,300 miles) per second.

Do you see how this fact provides a means by which a scientist (we will call him the observer) could calculate his own absolute motion? If light travels through a fixed, stationary ether with a certain speed, *c*, and if this velocity is independent of the velocity of its source, then the speed of light can be used as a kind of yardstick for measuring the observer's absolute motion. An observer moving in the same direction as a beam of light should find the beam passing him with a speed less than *c;* an observer moving toward a beam of light should find the beam approaching him with a velocity greater than *c*. In other words, measurements of the velocity of a beam of light should vary, depending on the observer's motion relative to the beam. These variations would indicate his true, absolute motion through the ether.

Physicists often describe this situation in terms of what they call an "ether wind." To understand just what they mean by this, consider again that moving train. We have seen how the speed of a man walking through the train at 4 kilometers per hour is always the same relative to the train, regardless of whether he walks toward the engine or toward the rear of the train. The same is true of the speed of sound waves inside a closed car. Sound is a wave

motion transmitted by molecules of air. Because the air is carried along by the car, sound will travel north in the car with the same velocity (relative to the car) with which it travels south.

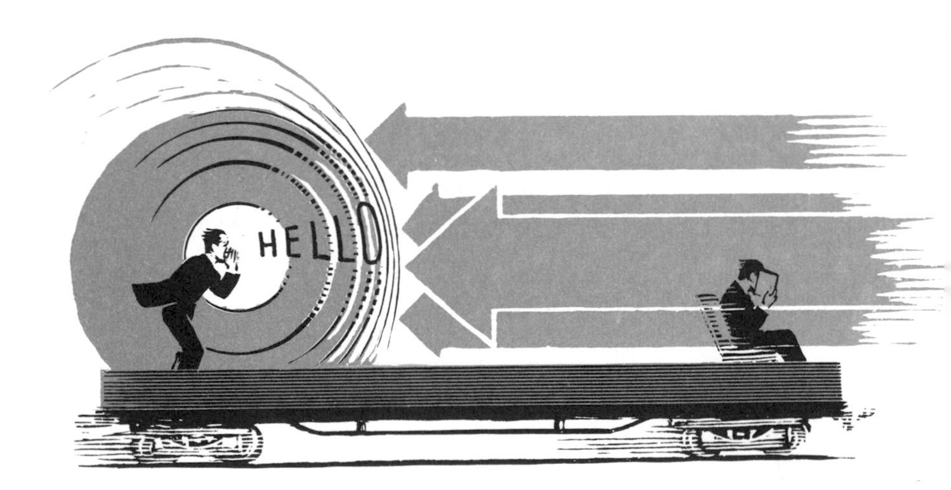

The situation alters if we move from the closed passenger car to an open flatcar. The air is no longer trapped inside the car. If the train moves at 100 kilometers per hour, there will be a wind of 100 kilometers per hour blowing back across the flatcar. Because of this wind, the speed of sound moving from the back to the front of the car will be less than normal. The speed of sound from front to back will be greater than normal.

Physicists of the nineteenth century believed that the ether surely must behave like the air that rushes over a moving flatcar. How could it be otherwise? If the ether is motionless, any object moving through it would have to encounter an "ether wind" blowing in the opposite direction. Light is a wave motion in this fixed ether. The velocity of light, measured on a moving object, would of course be influenced by such an ether wind.

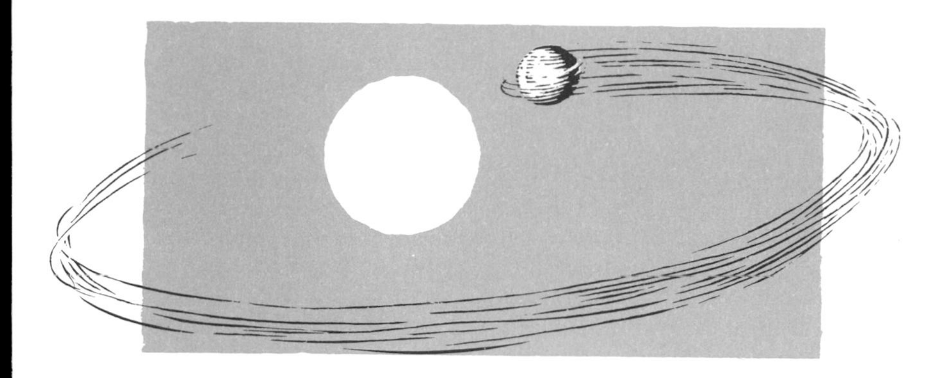

The earth is hurtling through space, on its trip around the sun, at a speed of about 30 kilometers per second. This motion, the physicists reasoned, should create an ether wind of 30 kilometers per second, blowing past the earth and through the spaces between its atoms. To measure the absolute motion of the earth—its motion with respect to the fixed ether—all that would be necessary would be to measure the speed of light as it travels back and forth in different directions on the earth's surface. Because of the ether wind, light would surely move faster in one direction than another. By comparing the various speeds of light as it is sent in different directions, it should then be possible to calculate the absolute direction and velocity of the earth's motion at any given instant. Such an experiment was first proposed in 1875, four years before Einstein was born, by the great Scottish physicist James Clerk Maxwell.*

* The suggestion appears in Maxwell's article on "Ether" in the ninth edition of *Encyclopædia Britannica.*

In 1881 Albert Abraham Michelson, then a young officer in the United States Navy, made just such an experiment. Michelson had been born in Germany, of Polish parents, but his father had taken him to America when he was two. After graduating from the U.S. Naval Academy at Annapolis and serving two years at sea, he became a teacher of physics and chemistry at the Academy. A leave of absence permitted him to study in Europe. It was at the University of Berlin, in the laboratory of the famous German physicist Hermann von Helmholtz, that young Michelson made his first attempt to detect an ether wind. To his great surprise, he could find no difference in the speed with which light traveled back and forth in any direction of the compass. It was as if a fish had discovered that it could swim in any direction through the sea without being able to detect the motion of water past its body; as if a pilot flying in the open cockpit of a plane could feel no wind against his face.

A distinguished Austrian physicist named Ernst Mach (we will hear more about him in Chapter 8) had for some time been criticizing the notion of absolute motion through the ether. He read Michelson's published report on the test and decided at once that the concept of an ether had to be discarded. However, most physicists refused to take this daring step. Michelson's apparatus had been crude. There was good reason to think that a better-designed experiment, with more sensitive equipment, would show positive results. Michelson himself thought so. He was disappointed in the "failure" of his test, and eager to try again.

Michelson resigned his naval commission to become a professor of physics at the Case School of Applied Science in Cleveland, Ohio. At nearby Western Reserve University, Edward Williams Morley was teaching chemistry. (The two schools are now merged as Case Western Reserve University.) The two men became good friends. "Outwardly," writes Bernard Jaffe in his book *Michelson and the Speed of Light,* "the two scientists were a study in contrast. . . . Michelson was good-looking and trim, always immaculately turned out. Morley, who was casual in dress, to say the least, fit the stereotype of the absent-minded professor. . . . He let his hair grow until it curled up on his shoulders, and he wore a great bristling red mustache that straggled almost to his ears."

In 1887, in Morley's basement laboratory, the two scientists made a second, more careful attempt to detect the elusive ether wind. Their experiment, which became known as the Michelson-Morley experiment, marked one of the great turning points in modern physics.

The apparatus was mounted on a square slab of stone about five feet on the side and more than a foot thick. The slab floated on liquid mercury. This eliminated vibrations, kept the slab horizontal, and permitted it to be rotated easily around a central pin. An arrangement of mirrors on the slab sent a light beam in a certain direction; then the mirrors reflected the beam back and forth in that same direction until it had made eight round trips. (This was done to make the path as long as possible and still keep the equipment on a device that could be rotated easily.) At the same time, the mirror arrangement sent a beam of light on eight round trips in a direction at right angles to the first beam.

The assumption was that when the slab was turned so that one beam traveled back and forth *parallel* to the ether wind, this beam would make the trip in a longer time than it would take the other beam to go the same distance *across* the wind. At first you might think the reverse would be true. Consider the light that travels with and against the wind. Would not the wind boost the speed by the same amount one way that it would retard the speed the other way? If so, the boosts and drags would cancel each other, and the time for the total trip would be the same as if there were no wind at all.

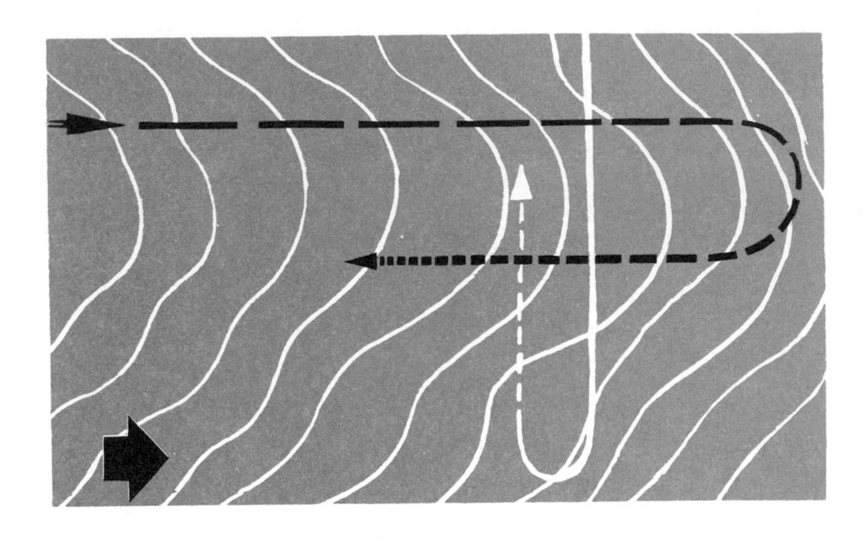

It is true that the wind would increase the velocity of light in one direction by the same amount that it would decrease the velocity in the other direction, but—and this is the crucial point—the wind would retard the speed for a *longer period of time*. Calculation quickly shows that the entire trip would take longer than if there were no wind. The wind would also have a retarding effect on the beam that traveled across the wind at right angles. This is also easily calculated. It turns out that this retarding effect is less than in the case of the beam traveling parallel to the wind.

There was little doubt, then, that *if* the earth moved through an immovable sea of ether, there would be an ether wind, and if there were an ether wind, the Michelson-Morley apparatus would detect it. In fact, both scientists were confident that they would not only find such a wind, but they could also

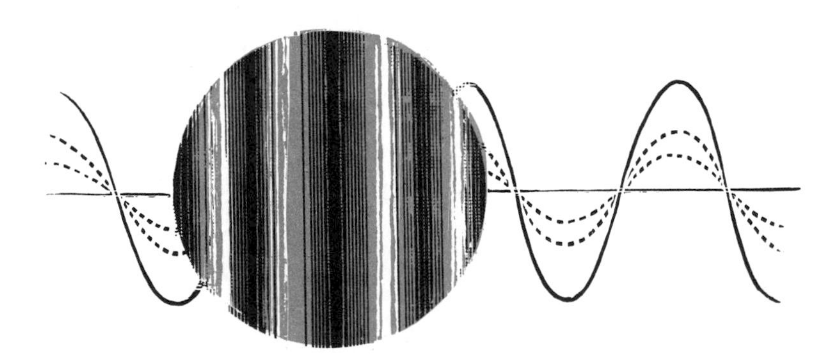

determine (by rotating the slab until there was a maximum difference in the time it took light to make the two journeys) the exact direction, at any given moment, of the earth's path through the ether.

It should be pointed out that the Michelson-Morley apparatus did not measure the actual velocities of each beam of light. The two beams, after making their respective back-and-forth trips, were combined into a single beam which was viewed through a small telescope. The apparatus would then be rotated slowly. Any alteration in the relative velocities of the two beams would cause a shifting of an interference fringe pattern of alternate light and dark bands.

Again Michelson was astounded and disappointed. This time the astonishment was felt by physicists all over the world. Regardless of how Michelson and Morley turned their apparatus, they found no sign of an ether wind! Never before in the history of science had the negative results of an experiment been so positive and so shattering. Michelson once more thought his experiment a failure. He never dreamed that this "failure" would make the experiment one of the most successful, revolutionary experiments in the history of science.

Later, Michelson and Morley repeated their test with even more accurate equipment. Other physicists did the same. An extremely accurate test was made in 1960 by Charles H. Townes of Columbia University. His apparatus, using a device called a maser (an "atomic clock" based on the vibrations of molecules), was so sensitive that he could have detected an ether wind even if the earth moved at a mere one-thousandth of its actual speed. There was no trace of such a wind.

Physicists at first were so amazed by the negative results of the Michelson-Morley test that they began inventing all sorts of explanations to save the ether-wind theory. Of course, if the experiment had been performed a few centuries earlier, as G. J. Whitrow points out in his book *The Structure and Evolution of the Universe*, a very simple explanation would immediately have occurred to everyone: the earth doesn't move! This theory seemed unlikely. The best explanation was a theory (much older than the first Michelson-Morley experiment) that the ether is dragged along by the earth, like air inside a closed train. This was Michelson's own guess. But other experiments, one by Michelson himself, ruled this out.

The strangest explanation of all was put forth by an Irish physicist, George Francis FitzGerald. Perhaps, he said, the ether wind puts pressure on a moving object, causing it to shrink a bit in the direction of motion. To determine the length of a moving object, its length at rest must be multiplied by the following simple formula, in which v^2 is the velocity of the object multiplied by itself, c^2 the velocity of light multiplied by itself:

$$\sqrt{1 - \frac{v^2}{c^2}}\,.$$

Study this formula and you will see that the amount of contraction is negligible at small velocities, increases as the velocity increases, and becomes great as the object's speed approaches the speed of light. Thus, a spaceship shaped like a long cigar would, if it moved with great speed, alter its shape to that of a short cigar. The speed of light is an unobtainable limit; when this is reached the formula becomes

$$\sqrt{1 - \frac{c^2}{c^2}},$$

which reduces to 0. Multiplying the length of the object by 0 results in 0. In other words, if an object could attain the speed of light, it would have no length at all in the direction of its motion!

FitzGerald's theory was put into elegant mathematical form by the Dutch physicist Hendrik Antoon Lorentz, who had independently thought of the same explanation. (Later, Lorentz became one of Einstein's closest friends, but at this time they did not know one another.) The theory came to be known as the Lorentz-FitzGerald (or the FitzGerald-Lorentz) contraction theory.

It is easy to understand how the contraction theory would explain the failure of the Michelson-Morley test. If the square slab and all the apparatus on it were contracted by a tiny amount in the direction in which the ether wind was blowing, the light would have a shorter total distance to travel. Even though the wind would have an overall drag effect on the beam's back-and-forth journey, the shorter path would permit the beam to finish the trip in the same time that it would take if there were no wind and no contraction. In other words, the contraction would be just enough to keep the speed of light a constant, regardless of the direction in which the Michelson-Morley apparatus is turned.

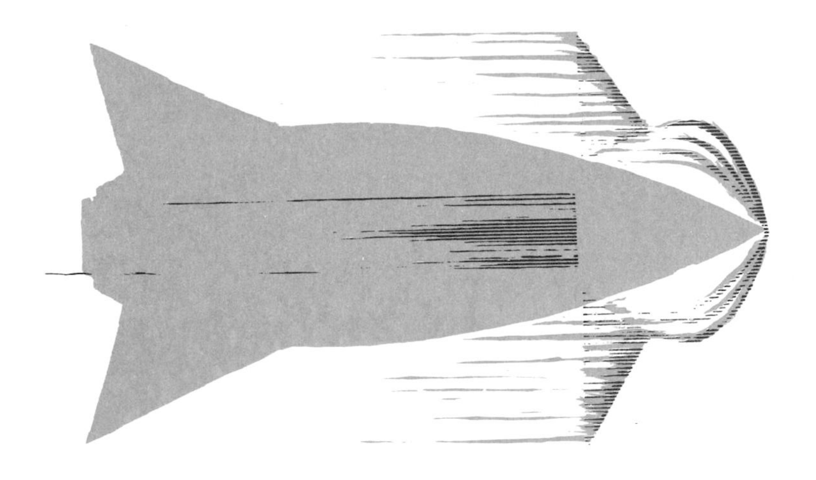

Why, you may ask, couldn't this theory be tested simply by measuring the length of the apparatus to see if it shortens in the direction of the earth's motion? The answer is that the ruler would shorten also, in the same proportion. As a result, measurements would come out the same as if there were no contraction. The contractions would apply to *everything* on the moving earth. The situation is similar to Poincaré's thought experiment (see Chapter 1) in which the cosmos suddenly grows a thousand times larger, except that in the Lorentz-FitzGerald theory the change would be in one direction only. Since the change applies to everything, there is no way to detect it. Within certain limits (the limits are set by topology—the study of properties that stay the same when an object is deformed), shape itself is as relative as size. The contraction of the apparatus, as well as the contraction of everything else on the earth, could be observed only by someone outside the earth and not moving with it.

Many writers on relativity have spoken of the Lorentz-FitzGerald contraction hypothesis as *ad hoc,* a Latin phrase (it rhymes with *sad sock*) meaning formulated "for this case alone," and incapable of being tested by any other experiment. This is not, as Adolf Grünbaum has pointed out, strictly true.

The contraction theory was *ad hoc* only in the sense that at the time there was no way to test it. In principle it is not at all *ad hoc*. In fact, it was definitely ruled out in 1932 by an important experiment called the Kennedy-Thorndike experiment.

Roy J. Kennedy and Edward M. Thorndike, two American physicists, repeated the Michelson-Morley test with this major difference: Instead of making the two arms of the apparatus as equal in length as possible, they made the lengths as different as possible. The apparatus was then rotated to see if there was any change in the difference between the times it took the two light beams to make round trips in the two directions. According to the contraction theory, this time difference would alter as the apparatus turned. It would be detected (as in Michelson's test) by changes in the interference fringes when the two beams were recombined. No such changes were observed. More accurate tests have been made in recent years by using a Mössbauer source of light (the Mössbauer effect will be discussed in Chapter 9) and a receiver mounted at opposite ends of a turntable which is then rotated rapidly. All such tests have falsified the contraction theory.

Although experiments of this sort could not be made in Lorentz's time, he realized that they could be made in principle, and there were good reasons to suppose that like Michelson's experiment, they would show negative results. To account for such probable results, Lorentz made an important addition to his original theory. He introduced changes in time. Clocks, he said, would be slowed down by an ether wind, and in just such a way as to make the velocity of light always measure 299,800 kilometers per second.

For one example of how this works out, suppose an attempt is made to measure the speed of light from A to B along a straight path in the direction the earth is moving. Two clocks at A are synchronized, then one clock is moved to B. A note is made of the time that a light beam starts from A and the time (measured by the other clock) that the beam is received at B. Since the light would be moving *against* the ether wind, its speed should be slowed down and the time of the trip should be a little longer than it would be if the earth were at rest. Do you see the flaw in this theory? The clock, in moving from A to B, also moves against the ether wind. This slows the clock at B down a bit, so that it is running slightly *behind* the clock at A. Result: the velocity of light still clocks at 299,800 kilometers per second.

The same thing happens (Lorentz maintained) if the speed of light is measured in the reverse direction, from B to A. Two clocks are synchronized at B, then one is taken to A. A light beam is sent from B to A, moving *with* the ether wind. The beam's speed is boosted by the wind, therefore the time taken by the light beam to make the trip should be a trifle less than if the earth were at rest. However, in moving the clock from B to A, it also went with the wind. The reduction of ether-wind pressure on the moving clock allowed the clock to gain a bit in time; therefore, when the experiment is made, the clock at A is running a bit *ahead* of the clock at B. Result: the velocity of light once again clocks at 299,800 kilometers per second.

Lorentz's new theory not only accounted for the negative results of the Michelson-Morley experiment; it also accounted for any conceivable experiment designed to detect changes in the speed of light as a result of an ether wind. Its equations for variations in length and time were worked out in such a way that every possible method of measuring the speed of light, from any frame of reference, would always give the same result. It is easy to understand why physicists were unhappy with this theory. It was *ad hoc* in the full sense of the word. It seemed little more than a weird effort to patch up the rents that had developed in the ether theory. There was no imaginable way either to confirm or refute it. Physicists found it hard to believe that if there were an ether wind, nature would go to such curious, drastic, almost prankish lengths to prevent it from being detected. Arthur Stanley Eddington, a distinguished British astronomer who was one of Einstein's earliest admirers, described the situation aptly by quoting the following lines from Lewis Carroll's song of the White Knight in *Through the Looking-Glass:*

> But I was thinking of a plan
> To dye one's whiskers green,
> And always use so large a fan
> That they could not be seen.

Lorentz's new theory, with its time as well as length changes, seemed almost as absurd as the White Knight's plan. But try as they would, physicists were unable to think of a better plan.

The next chapter will show how Einstein's special theory of relativity pointed to a bold, remarkable way out of this extraordinary confusion.

1905

3 The Special Theory of Relativity, Part I

In 1905, when Albert Einstein published his famous paper on what later became known as the special theory of relativity, he was a young married man of twenty-six, working as an examiner for the Swiss patent office. His career as a physics student, at The Polytechnic Institute of Zürich, had not been impressive. He had preferred to read, think, and dream on his own rather than cram his mind with unessential facts in order to pass examinations with high marks. He tried teaching physics, but he was a clumsy teacher and lost several such positions.

There is another side to this history. From the time that he was a small boy, Einstein had thought deeply about the fundamental laws of nature. He later recalled the two greatest "wonders" of his childhood: a compass his father showed him when he was four or five, and a book on Euclidian geometry that he read when he was twelve. These two "wonders" are symbolic of Einstein's life work: the compass a symbol of physical geometry, the structure of that "huge world" outside of us, about which we can never be absolutely certain; the book a symbol of pure geometry, a structure that is absolutely certain but independent of the actual world. Before he was sixteen Einstein had acquired, largely by his own efforts, a solid understanding of basic mathematics, including analytic geometry and calculus.

While Einstein was working in the Swiss patent office he was reading and thinking about all sorts of perplexing problems that had to do with light and motion. His special theory was a brilliant attempt to account for a wide variety of unexplained experiments, of which the Michelson-Morley test had been the most startling and best publicized. It is important to understand that there were many other experiments that had created a highly unsatisfactory state of affairs with respect to theory about electromagnetic phenomena. If the Michelson-Morley test had never been made, the special theory would still have been formulated. Einstein himself later spoke about the small role that it actually played in his thinking. Of course, if Michelson and Morley had detected an ether wind, the special theory would have been ruled out from the start. But the negative result of the test was only one of many things that led Einstein to his theory.

We have seen how Lorentz and FitzGerald had tried to save the ether-wind theory by assuming that the pressure of the wind, in some not-yet-understood way, causes an actual physical contraction of objects in motion. Einstein, following the footsteps of Ernst Mach, took a bolder view. The reason Michelson and Morley were unable to detect an ether wind, Einstein said, is simple: *There is no ether wind.* He did not say that there is no ether; only that the ether, if it exists, is of no value in measuring uniform motion. (In recent years a number of prominent physicists have proposed that the term "ether" be restored, though not, of course, in the old sense of an immovable frame of reference.)

Classical physics—the physics of Isaac Newton—made clear that if you are on a uniformly moving object, say a train car that is closed on all sides so you cannot see the scenery go by, there is no mechanical experiment by which you can prove that you are moving. (This assumes, of course, that the uniform motion is completely smooth, with no bumps or swaying of the car that can serve as clues to motion.) If you toss a ball straight up in the air, it comes straight down again. This is exactly what would happen if the train were standing still. An observer on the ground outside the moving car, if he could see through the sides of the car, would see the ball's path as a curve. But to you inside the car, the ball goes straight up and down. It is fortunate

that objects behave in this way. Otherwise one could never play a game like tennis or baseball. Each time the ball went up in the air, the earth would move out from under it.

The special theory of relativity carries the classical relativity of Newton forward another step. It says that in addition to being unable to detect the train's motion by a *mechanical* experiment, it also is impossible to detect its motion by an *optical* experiment, more precisely, by an experiment with electromagnetic radiation. The special theory can be put in a nutshell: It is not possible to measure uniform motion in any absolute way. If we are on a smoothly, uniformly moving train, we have to peek through a window and look at some other object, say a telephone pole, to make sure we are moving. Even then we cannot say positively whether the train is moving past the pole or the pole moving past the train. The best we can do is say that the train and the ground are in relative uniform motion.

Note the constant repetition in the last paragraph of that word "uniform." Uniform motion is motion in a straight line at a constant speed. Nonuniform or *accelerated* motion is motion that is getting faster or slower (when it is getting slower the acceleration is said to be negative), or motion along a path that is not a straight line. The special theory of relativity has nothing new to say about accelerated motion.

The relativity of uniform motion seems harmless enough, but the fact is that it plunges us immediately into a strange new world that at first seems to resemble nothing so much as the nonsense world behind Lewis Carroll's looking-glass. For if there is no way to measure uniform motion relative to a universal, fixed frame of reference like the ether, then light must behave in an utterly fantastic way, completely contrary to all experience.

Consider an astronaut in a spaceship that is racing alongside a light beam. The ship is traveling with half the speed of light. The astronaut will find, if he makes the proper measurements, that the beam is still passing him at its usual velocity of 299,800 kilometers per second! Think about this for a moment and you will soon realize that this must indeed be the case if the notion of an ether wind is discarded. If the astronaut found that light slowed down relative to his motion, he would have detected the very ether wind that Michelson and Morley failed to detect. Similarly, if his spaceship travels directly toward a source of light, moving with half the speed of light, will he find the beam approaching him twice as fast? No, it is still moving toward him at 299,800 kilometers per second. Regardless of how he moves relative to the beam, his measurements will always give the beam the same speed.

Frequently one hears the remark that relativity theory makes everything in physics relative, that it destroys all absolutes. Nothing could be further from the truth. It makes some things relative that were previously thought absolute, but in doing so, it introduces new absolutes. In classical physics the speed of light was relative in the sense that it should appear to change depending on the motion of the observer. In the special theory of relativity, the speed of light becomes, in this sense, a new absolute. No matter how a source of light moves, or how an observer moves, the speed of light relative to the observer never changes.

Imagine two spaceships, A and B. There is nothing in the cosmos except these two ships. They move toward each other at uniform speed. Is there any way that astronauts on either ship can decide which of the following three situations is "true" or "absolute"?

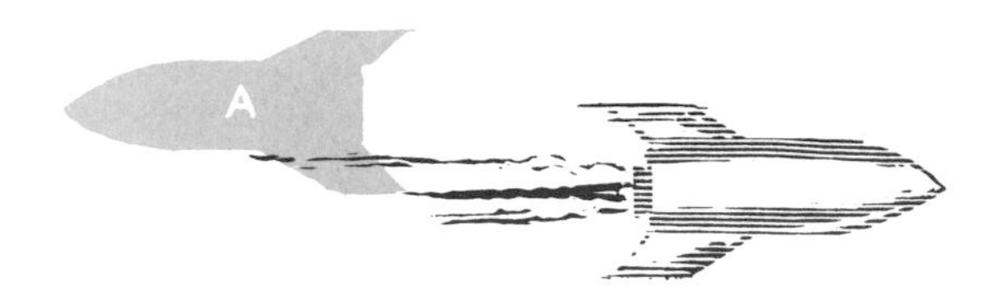

1. Spaceship A is at rest, B is moving.

2. Spaceship B is at rest, A is moving.

3. Both ships are moving.

Einstein's answer is: No, there is no way to decide. An astronaut on either ship can, if he wishes, choose to make ship A the fixed frame of reference. There is no experiment of any sort, including experiments with light or any other electrical or magnetic phenomena, that will prove this choice wrong. The same is true if he chooses to make ship B the frame of reference. If he prefers to regard both ships as moving, he simply chooses a frame of reference outside the two ships; a spot relative to which both ships are in motion. There is no question of one of these choices being "right" and the others "wrong." To speak of an absolute motion of either ship is to say something that has no meaning. There is only one reality: a relative motion that brings the ships closer together at uniform speed.

In a book of this sort it is impossible to go into technical details about the special theory, especially details that involve its mathematics. We must be content with mentioning some of the more surprising consequences that follow logically from what Einstein, in his first paper on relativity, calls the two "fundamental postulates" of his theory:

1. There is no way to tell whether an object is at rest or in uniform motion relative to a fixed ether.

2. Regardless of the motion of its source, light always moves through empty space with the same constant speed.

(The second postulate should not be confused, as it so often is, with the constant speed of light relative to a uniformly moving *observer*. This is a *deduction* from the postulates. Note also that we are concerned only with light's speed in a vacuum. Light travels more slowly through light-transmitting substances such as air or glass, otherwise no lens would refract light.)

Other physicists had, of course, considered these two postulates. Lorentz had tried to reconcile them by his theory in which absolute lengths and times were altered by the pressure of the ether wind. Most physicists thought this too radical a violation of common sense. They preferred to believe that the postulates were incompatible and at least one of them must be wrong. Einstein thought about the problem more deeply. The postulates were incompatible, he said, only if one refused to give up the classical view that length and time were absolute. When Einstein published his theory he did not know that Lorentz had been thinking along similar lines, but like Lorentz he recognized that measurements of length and time must depend on the relative motion of object and observer. Lorentz, however, had gone only halfway. He had kept the notion of an absolute length and time for objects at rest. He thought that the ether wind distorted "true" length and time. Einstein went the full way. There is, he said, no ether wind. There is no meaning to the concepts of absolute length and time. This is the key to Einstein's special theory. When he turned it, all sorts of locks began slowly to open.

To explain his special theory in a nontechnical way, Einstein once introduced the following famous thought experiment. Imagine, he said, an observer M who is standing beside a railroad track. At a certain distance down the track is a spot A. At the same distance up the track is a spot called B. Lightning happens to strike simultaneously at points A and B. The observer knows these events are simultaneous, because he sees the two flashes at the same instant. Since he is midway between them, and since light travels at a constant speed, he calculates that the lightning struck simultaneously in the two spots.

Now assume that when the lightning strikes, a train is moving at great speed along the track in the direction from A to B. At the instant the two flashes occur, an observer on the train—we call him M′—is exactly opposite observer M on the track. Since M′ is moving toward one flash and away from the other, he will see the flash at B before he sees the flash at A. Knowing that he is in motion, he will make allowances for the speed of light; he, too, will calculate that the two flashes occurred simultaneously.

All well and good. But according to the two fundamental postulates of the special theory (and confirmed by the Michelson-Morley test), we have just as much right to assume that the train is at rest while the ground moves rapidly backward under the train's wheels. From *this* point of view, M′, the observer on the train, will conclude that the flash at B actually did occur ahead of the flash at A, just as he observed them. He knows that he is midway between the two flashes and, since he regards himself as at rest, he is forced to conclude that the flash he saw first must have occurred before the flash he saw second.

M, the observer on the ground, is forced to agree. True, he sees the flashes as simultaneous, but now *he* is the one who is assumed to be moving. When he makes allowances for the speed of light and the fact that he is moving toward the flash at A and away from the flash at B, he will calculate that the flash at B must have taken place first.

We are driven to conclude, therefore, that the question of whether the flashes are simultaneous cannot be answered in any absolute way. The answer depends on the choice of a frame of reference. Of course, if two events occur simultaneously *at the same spot*, it can be said absolutely that they are simultaneous. When two airplanes collide in midair, there is no frame of reference from which the smashing of both planes will not be simultaneous. But the greater the distance between two events, the greater the difficulty of deciding about simultaneity. It is important to understand that this is not just a question of being unable to learn the truth of the matter. *There is no actual truth of*

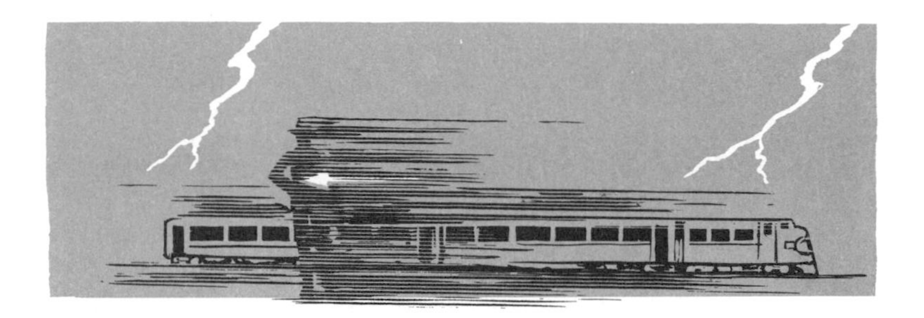

the matter. There is no absolute time throughout the universe by which absolute simultaneity can be measured. Absolute simultaneity of distant events is a meaningless concept.

How radical this notion is can be seen by a thought experiment in which vast distances and enormous speeds are involved. Suppose that someone on Planet X, in another part of our galaxy, is trying to communicate with the earth. He sends out a radio message. This is, of course, an electromagnetic wave that travels through space with the speed of light. Assume that the earth and Planet X are ten light-years apart, which means that it takes ten years for the message to travel to the earth. Twelve years before a radio astronomer on earth receives the message, the astronomer had received a Nobel Prize. The special theory permits us to say, without qualification, that he received this prize *before* the message was sent from Planet X.

Ten minutes after receiving the message, the astronomer sneezes. The special theory also permits us to say, without qualification, and for all observers in any frame of reference, that the astronomer sneezed *after* the message was sent from Planet X.

Now suppose that sometime during the ten-year period, while the radio message was on its way to the earth (say, three years before the message was received), the astronomer fell off his radio telescope and broke a leg. The special theory does *not* permit us to say without qualification that he broke his leg before or after the sending of the message from Planet X.

The reason is this: one observer, leaving Planet X at the time the message is sent and traveling to the earth with a speed judged from the earth to be slow, will find (according to his measurements of the passing of time) that the astronomer broke his leg *after* the message was sent. Of course, he will arrive on earth long after the message is received, perhaps centuries after. But when he calculates the date on which the message was sent, according to his clock it will be earlier than the date on which the astronomer broke his leg. On the other hand, another observer, who also leaves Planet X at the time the message is sent, but who travels very close to the speed of light, will find that the astronomer broke his leg *before* the message was sent. Instead of taking

centuries to make the trip, he will make it in, say, only a trifle more than ten years as calculated on the earth. But because of the slowing down of time on the fast-moving spaceship, it will seem to the ship's astronaut that he made the trip in only a few months. He will be told on the earth that the astronomer broke his leg a little more than three years before. According to the astronaut's clock, the message was sent a few months before. He will conclude that the leg was broken years before the message left Planet X.

If the astronaut traveled as fast as light (of course this is purely hypothetical, not possible in fact), his clock would stop completely. It would seem to him that he made the trip in zero time. From his point of view the two events, the sending of the message and its reception, would be simultaneous. *All* events on earth during the ten-year period would appear to him to have occurred before the message was sent. Now, according to the special theory there is no "preferred" frame of reference: no reason to prefer the point of view of one observer rather than another. The calculations made by the fast-moving astronaut are just as legitimate, just as "true," as the calculations made by the slow-moving astronaut. There is no universal, absolute time that can be appealed to for settling the differences between them. The instant "now" has meaning only for the spot you occupy. You cannot assume that a "now" exists simultaneously for all spots of the universe.

This breakdown in the classical notion of absolute simultaneity is by all odds the most "beautifully unexpected" aspect of the special theory. (The phrase "beautifully unexpected" is from a speech on relativity by the nuclear physicist Edward Teller.*) Newton took for granted that one universal time permeated the entire cosmos. So did Lorentz and Poincaré. It was *this* that prevented them from discovering the special theory ahead of Einstein! Einstein had the genius to see that the theory could not be formulated in a comprehensive, logically consistent way without giving up completely the notion of a universal cosmic time.

There are, said Einstein, only local times. On the earth, for example, everyone is being carried along through space at the same speed; therefore, their watches all run on the same "earth time." A local time of this sort, for a moving object like the earth, is called that object's "proper time." There is still an absolute "before" and "after" (obviously no astronaut is going to die before he is born), but when events are separated by vast distances, there are long time intervals within which it is not possible to say which of two events is before or after the other. The answer depends on the observer's motion with respect to the two events. The decision reached by one observer is just as "true" as a different decision reached by another observer. All this follows with iron logic from the two fundamental postulates of the special theory.

* The speech was reprinted as "The Geometry of Space and Time," in *The Mathematics Teacher* (November 1961).

When the concept of simultaneity falls, other concepts fall with it. Time becomes relative because observers differ in their estimates of the time that elapses between the same two events. Length also becomes relative. The length of a moving train cannot be measured without knowing exactly where the front and back ends are *at the same instant.* If someone reports that at 1:00 o'clock the front end of a train was exactly opposite him and that the back end was a mile down the track at some time between 12:59 and 1:01, there obviously is no way of determining the exact length of the train. In other words, a method of establishing exact simultaneity is essential for the accurate measurements of distances and the lengths of moving objects. In the absence of such a method, the lengths of moving objects become dependent on the choice of a frame of reference.

For example, if two spaceships are in relative motion, an observer on each ship will measure the other ship as contracted slightly in the direction of its motion. At ordinary speeds this change is extremely minute. The earth, which moves at 30 kilometers per second around the sun, would appear, to an observer at rest relative to the sun, as shortened only by a few inches. When relative speeds are very great, however, the change becomes significant. It turned out, happily, that the same formula for contraction that had been devised by FitzGerald and Lorentz, to explain the Michelson-Morley test, could be applied here. In relativity theory it is still called the

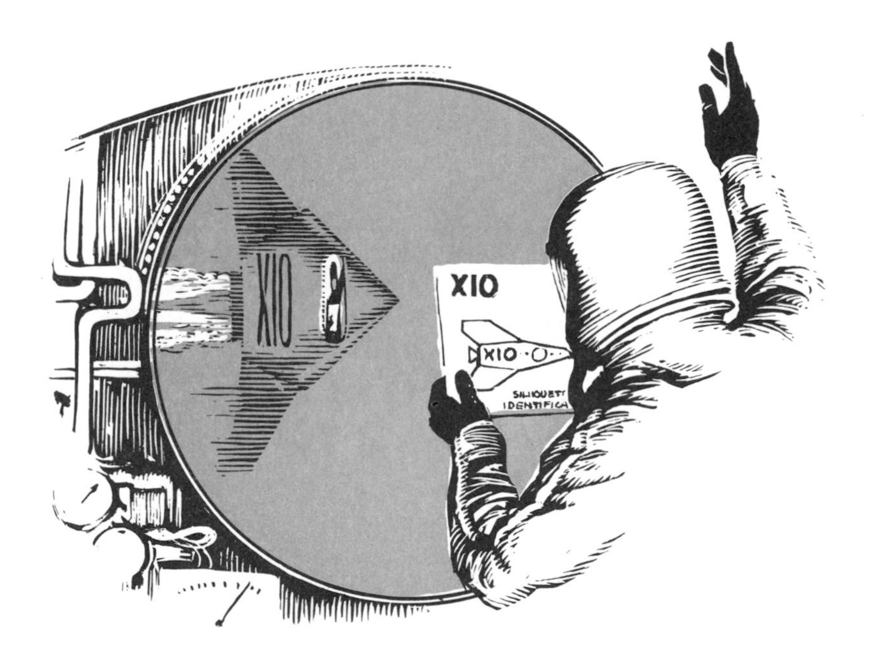

Lorentz-FitzGerald contraction, but it would be less confusing if it had some other name because Einstein gave the formula a fundamentally different interpretation.

For Lorentz and FitzGerald the contraction was a physical change, caused by pressure of the ether wind. For Einstein it had only to do with the results of measurement: in this case, when the astronauts on one spaceship measure the length of the other ship. The observers on each ship detect no change in the length of their own ship, or the lengths of objects inside it. When they measure the other ship, however, they find it shorter. Lorentz and FitzGerald still thought of moving objects as having absolute "rest lengths." When objects contracted, they were no longer their "true" lengths. Einstein, by giving up the ether, made the concept of absolute length meaningless. What remained was *length as measured*, and this turned out to vary with the relative speed of object and observer.

How is it possible, you ask, for each ship to be shorter than the other? You ask an improper question. The theory does not say that each ship is shorter than the other; it says that astronauts on each ship *measure* the other ship as shorter. This is a quite different matter. If two people stand on opposite sides of a huge concave lens, each sees the other as smaller, but that is not the same as saying that each *is* smaller.

In addition to apparent changes in length, there are also apparent changes in time. Astronauts on each ship will find that clocks* on the other ship are running slower. A simple thought experiment shows that this must indeed be the case. Imagine that you are looking out through the porthole of one

* The word "clock" is used here and throughout the book for any type of periodic process that is not dependent on gravity: the movement of a balance-wheel clock, the beating of a heart, and so on. It is good to bear in mind that gravity clocks, such as pendulum clocks and sand glasses, would be useless under the described conditions.

spaceship into the porthole of another ship. The two ships are passing each other with a uniform speed close to that of light. As they pass, a beam of light on the other ship is sent from its ceiling to its floor. There it strikes a mirror and is reflected back to the ceiling again. You will see the path of this light as a V. If you had sufficiently accurate instruments (of course no such instruments exist), you could clock the time it takes this light beam to traverse the V-shaped path. By dividing the length of the path by the time, you obtain the speed of light.

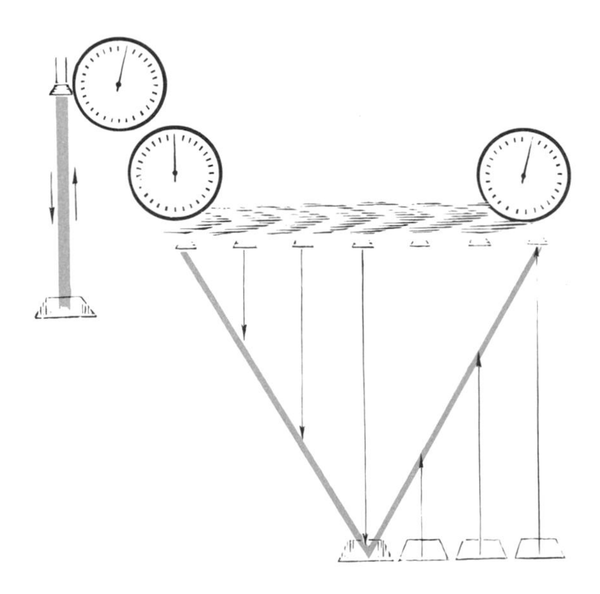

Now suppose that while you clock the light beam on its V-shaped path, an astronaut inside the other ship is doing the same thing. From his point of view, assuming *his* ship to be the fixed frame of reference, the light simply goes down and up along the same line, obviously a shorter distance than along the V that you observed. When he divides this distance by the time it took the beam to go down and up, he also obtains the speed of light. Because the speed of light is constant for all observers, he must get exactly the same final result that you did: 299,800 kilometers per second. But his light path is shorter. How can his result be the same? There is only one possible explanation: his clock is slower. Of course, the situation is perfectly symmetrical. If you send a beam down and up inside your ship, he will see its path as V-shaped. He will deduce that *your* clock is slower.

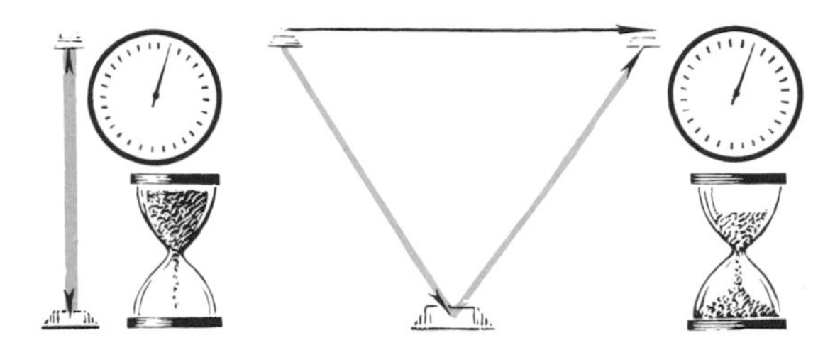

The fact that these bewildering changes of length and time are called "apparent" does not mean that there is a "true" length or time which merely "appears" different to different observers. Length and time are relative concepts. They have no meaning apart from the relation of an object to an observer. There is no question of one set of measurements being "true," another set "false." Each is true relative to the observer making the measurements; relative to his frame of reference. *There is no way that measurements can be any truer.* In no sense are they optical illusions, to be explained by a psychologist. They can be recorded on instruments. They do not require a *living* observer.

Mass, too, is a relative concept, but we must defer this and other matters to the next chapter.

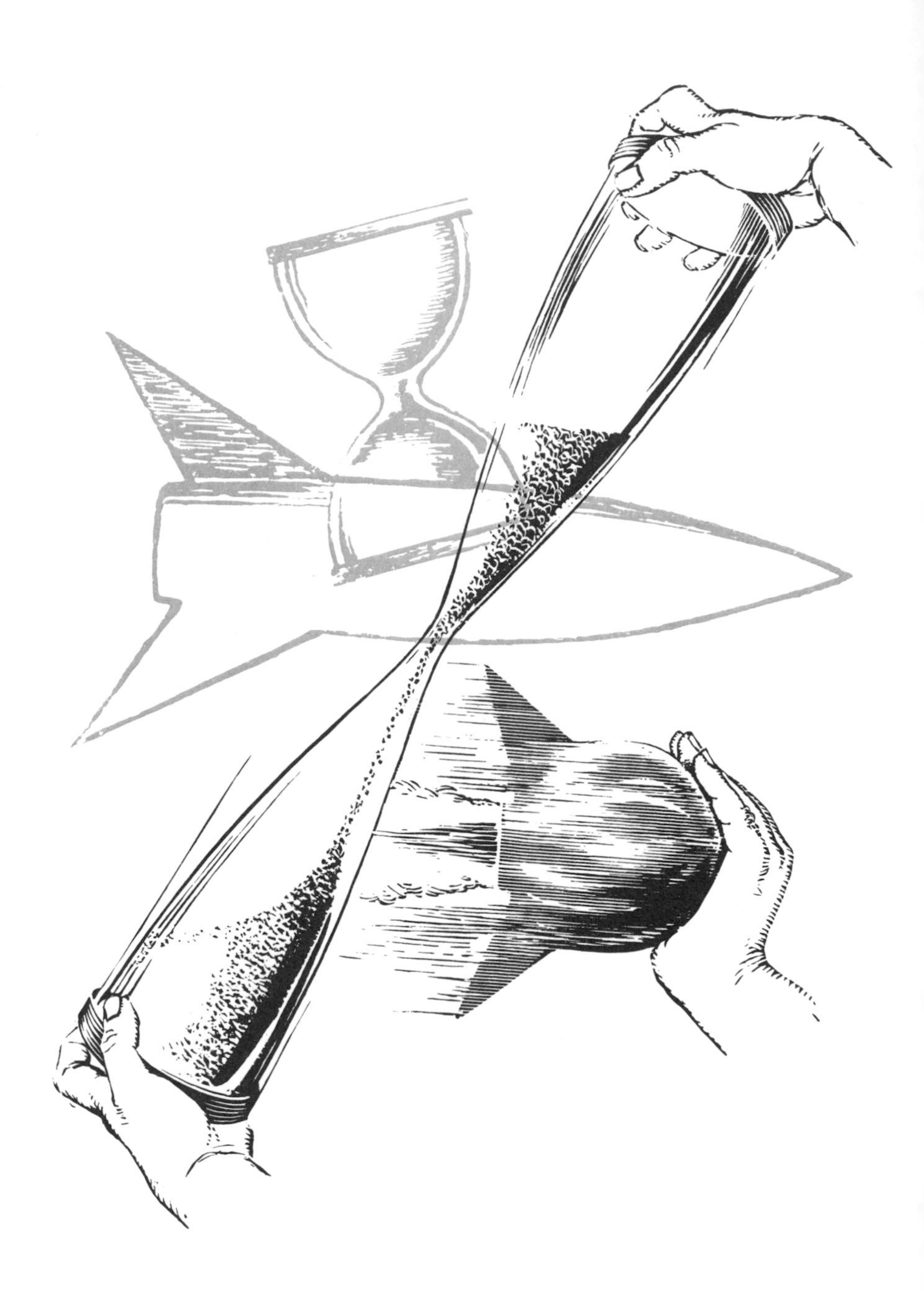

4

The Special Theory of Relativity, Part II

Length and time, as was shown in the previous chapter, are relative concepts. If two spaceships pass each other with uniform velocity, observers on each ship will find that astronauts on the other ship are thinner and moving about more slowly. If the relative speed is great enough, they will seem to move like actors in a slow-motion picture. All phenomena with periodic movements will seem reduced in speed: tuning forks, balance-wheel watches, heartbeats, vibrating atoms, and so on. As Eddington once expressed it, even cigars on the other ship will seem to last longer. A six-foot astronaut, standing erect in a horizontally moving ship, will still appear six feet tall, but his body will seem thinner in the line of travel. When he lies down with his body in line with the ship's motion, his body will be restored to normal width but he will now seem shorter from head to toes.

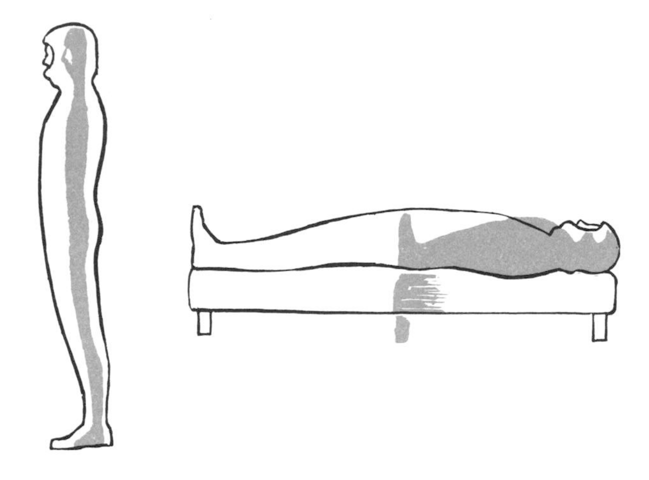

If two spaceships actually could pass each other with a relative speed great enough to make such changes significant, all sorts of technical difficulties would make it virtually impossible for observers on either ship to *see* such changes. Writers like to explain relativity by using oversimplified dramatic illustrations. These colorful illustrations do not describe changes that actually could be observed, either by the human eye or by any instruments presently known. They should be thought of as changes that could, in principle, be inferred by the astronauts on the basis of measurements, with sufficiently precise instruments and after making necessary corrections for the velocity of light. Whenever we speak of an "observer" we mean an imaginary, idealized person, attached to a specified reference frame, who reaches certain conclusions based on his measuring instruments.*

In addition to changes in length and time, there are also relativistic changes in mass. Mass, in a rough sense, is a measure of the amount of matter in an object. A lead ball and a cork ball may be the same size, but the lead ball is more massive. It contains a greater concentration of matter.

* The actual appearance of an object—how it would look on a photograph taken instantaneously—when observer and object are passing each other at high relativistic speeds, is a complicated matter that has been investigated only since 1959. Classical laws of optics combine with Lorentz contractions to produce surprising results. A sphere, for example, always appears as a circular disk. Under certain conditions a cube seems to have been rotated. The interested reader will find such matters discussed in James Terrell, "Invisibility of the Lorentz Contraction," *Physical Review* (November 15, 1959); V. F. Weisskopf, *Physics Today* (September 1960); G. D. Scott and M. R. Viner, "The Geometrical Appearance of Large Objects Moving at Relativistic Speeds," *American Journal of Physics* (July 1965).

There are two ways to measure an object's mass. It can be weighed or it can be determined how much force is needed to accelerate the object by a certain amount. The first method is not a very good one because the results vary with the local strength of gravity. A lead ball carried to the top of a high mountain will weigh a trifle less than before, although its mass remains exactly the same. On the moon its weight would be considerably less than on the earth. On Jupiter its weight would be considerably more.

The second method of measuring mass gives the same result regardless of whether one is on the earth, the moon, or Jupiter, but it is subject to a different and odder kind of variation. To determine the mass of a moving object by this method, one must measure the force required to accelerate it by a certain amount. Clearly, a stronger push is needed to start a cannonball rolling than to start a cork ball rolling. Mass measured in this way is called *inertial mass* to distinguish it from *gravitational mass*. Measurements of this nature cannot be made without making measurements of time and distance. The inertial mass of a cannonball, for example, is expressed by the amount of force required to increase its speed (distance per unit time) by so much per unit of time. As we have seen, time and distance measurements vary with the relative speed of object and observer. As a result, measurements of inertial mass also vary.

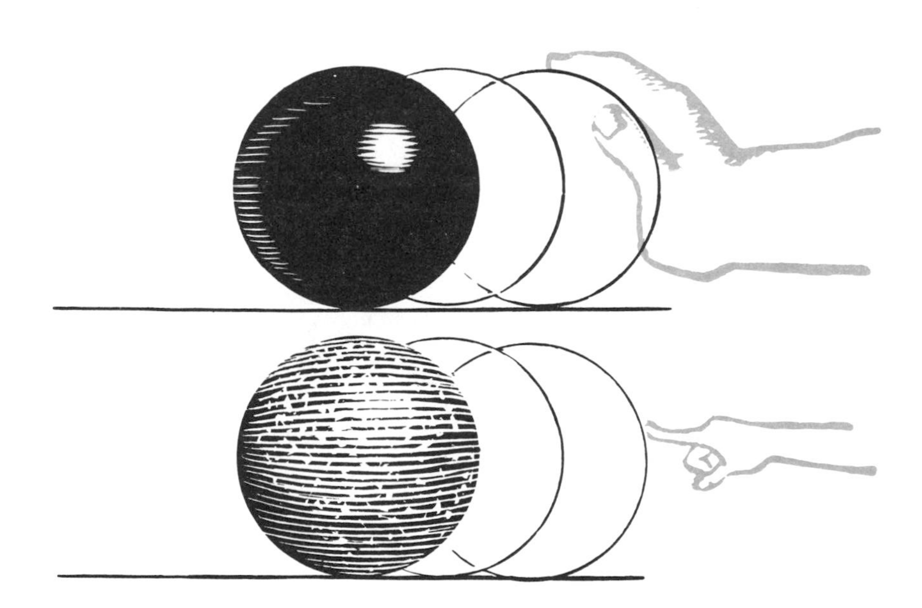

In Chapter 5 we will return to the concept of gravitational mass and its relation to inertial mass. Here we are concerned only with inertial mass as measured by an observer. For observers at rest relative to an object—for example, astronauts carrying an elephant on a spaceship—the object's inertial mass remains the same regardless of the speed with which the ship is traveling. The elephant's mass, as measured by such observers, is called its *proper mass* or *rest mass*. The same elephant's inertial mass, measured by an observer in relative motion with the elephant (for example, by an observer on the earth), is called the elephant's *relativistic mass*. The rest mass of an object never varies. Its relativistic mass does. Both are measurements of inertial mass. In this chapter we are concerned only with inertial mass; when the word "mass" is used it should always be taken in that sense.

All three variables—length, time, mass—are covered by the same Lorentz contraction formula given in Chapter 2. Length and the rate of clocks vary in the same direct proportion, so the formula is the same for each. Mass and the length of time intervals vary in inverse proportion, which means that the formula has to be written like this:

$$\frac{1}{\sqrt{1-\frac{v^2}{c^2}}}.$$

The mass of an object, measured by an observer in uniform motion relative to the object, is obtained by multiplying the object's rest mass by the above formula (where v is the relative velocity of the object, c the speed of light).

For example, if the relative speed of two spaceships is close to 259,635 kilometers per second, observers on either ship will find the other ship half as long, its clocks running half as fast, its hours twice as long, and its mass twice as large. Of course, the astronauts will find everything completely normal inside their own ship. If the ships could attain a relative speed equal to that of light, observers on each ship would think the other ship had shrunk to zero in length, acquired an infinite mass, and that time on the other ship had slowed to a full stop!

If inertial mass did not vary in this way, then the steady application of force, such as the force supplied by rocket motors, could keep increasing a ship's velocity until it passed the speed of light. This cannot occur because as the ship goes faster and faster (from the standpoint, say, of an observer on the earth), its relativistic mass keeps increasing in the same proportion as its length and time are decreasing. When the ship has contracted to one-tenth its rest length, its relativistic mass has become ten times as great. It is offering ten times as much resistance to its rocket motors; therefore, ten times as much force is required to produce the same increase in speed as would be required if the ship were at rest. The speed of light can never be reached. If

it were reached, the outside observer would find that the ship had shrunk to zero length, had acquired an infinite mass, and was exerting an infinite force with its rocket motors. Astronauts inside the ship would observe no changes in themselves, but they would find the cosmos hurtling backward with the speed of light, cosmic time at a standstill, every star flattened to a disk and infinitely massive.

Only a science-fiction writer would dare speculate on what astronauts might observe from a ship moving faster than light. Perhaps the cosmos would appear to turn inside out and become its own mirror image, stars would acquire negative mass, and cosmic time would run backward. I hasten to add that none of this follows from the formulas of the special theory. If the speed of light is exceeded, the formulas give values to length, time, and mass that are what mathematicians call "imaginary numbers"—numbers that involve the square root of minus one. Who can say? Maybe a ship that broke the light barrier would plunge straight into the Land of Oz!

After learning that nothing can outrun light, beginning students of relativity are often perplexed when they come across references to velocities faster than light. To understand exactly what relativity has to say on this point, it will be best to introduce the term "inertial frame." (Earlier writers on relativity called it "inertial system" or "Galilean system.") When an object like a spaceship is in uniform motion, that object and all objects moving along with it in the same direction and with the same speed (such as all the objects inside the ship) are said to be attached to the same inertial frame. (To be more technical, the inertial frame is the Cartesian coordinate system to which the spaceship is attached.) Outside the context of a specific inertial frame, the special theory no longer applies and there are many ways that speeds faster than light can be observed.

Consider, for example, this simple situation. A spaceship, traveling at three-fourths the speed of light, passes overhead going due east. At the same instant another spaceship, also traveling at three-fourths the speed of light, passes overhead going due west. From your frame of reference, attached to the inertial frame of the earth, the two ships pass each other with a relative velocity of one and one-half times the speed of light. They approach at that speed, move apart at that speed. There is nothing in relativity theory to deny this. However, the special theory does insist that *if you were riding on either ship*, you would calculate the relative speed of the ships to be less than that of light.

In this book we are trying our best to avoid the mathematics of relativity, but like the Lorentz contraction formula, the formula below is too simple to leave out. If x is the speed of one ship relative to the earth and y is the speed of the other ship relative to the earth, then the speed of the ships relative to each other, *as seen from the earth*, is, of course, x plus y. But as seen by an observer on either ship, we have to add velocities by the following formula:

$$\frac{x+y}{1+\frac{xy}{c^2}}.$$

In this formula, c is the velocity of light. It is easy to see that when the speeds of the ships are small compared to light, the formula gives a result that is almost the same as the result obtained by adding the two velocities in the usual manner. But if the speeds of the ships are very great, the formula gives a quite different result. Take the limiting case and assume that instead of spaceships there are two beams of light passing overhead in opposite directions. The earth observer sees them separate with a speed of $2c$, or twice the speed of light. But if he were riding on one beam, he would calculate this speed, according to the formula, as

$$\frac{c+c}{1+\frac{c^2}{c^2}},$$

which, of course, reduces to the value of c. In other words, he would see the other beam moving away from him with the speed of light.

Suppose that a beam of light passes overhead at the same time that a spaceship moves in the opposite direction with a speed of x. From the earth's inertial frame, ship and light pass each other with a speed of c plus x. The reader may enjoy using the formula to calculate the speed of light as observed from the spaceship's inertial frame. It turns out, of course, to be c again.

Outside the province of the special theory, which is concerned only with inertial frames, it is still possible to speak of the speed of light as an absolute limit. But now it has to be phrased in a different way: There is no way to send a *signal*, from one material body to another, with a speed faster than light. "Signal" is here used in a wide sense to include any sort of cause-and-effect chain by which a message can be transmitted: the sending of a physical object, for instance, or the transmission of any type of energy such as a sound wave, electromagnetic wave, shock wave in a solid, and so on. A message cannot be sent to Mars with a speed greater than the speed of light. This cannot be done by writing a letter and sending it in a rocket, because as we have seen, the rocket's relative speed must always be less than the speed of light. If the message is coded and sent by radio or radar, it goes *at* the speed of light. No other type of energy can provide a faster transmission of the code.

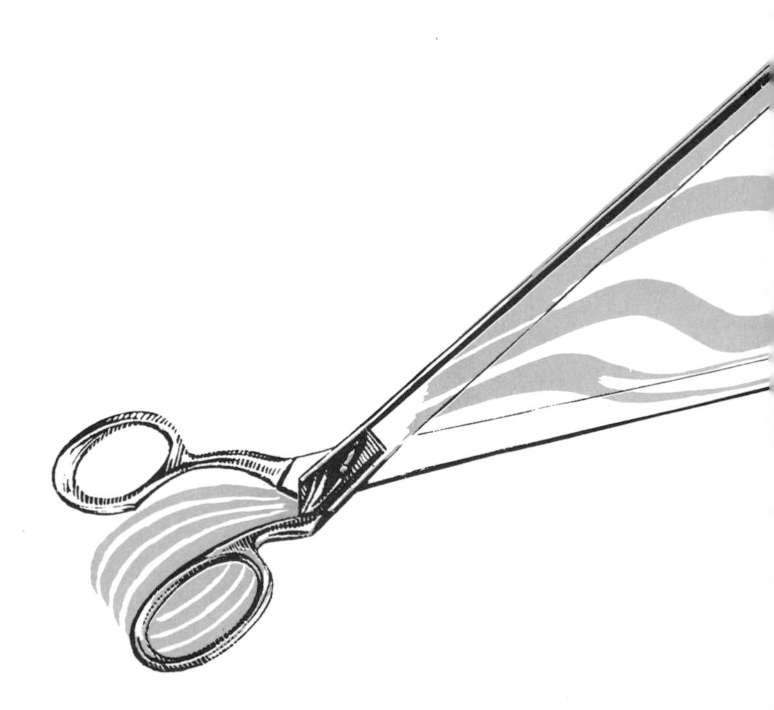

Although signals cannot be sent faster than the speed of light, it is possible to observe certain types of motion that, relative to the observer, will have a speed faster than light.* Imagine a gigantic pair of scissors, the blades as long as from here to the planet Neptune. The scissors begin to close with uniform speed. As this happens, the point where the cutting edges intersect will move toward the points of the scissors with greater and greater velocity. Imagine yourself sitting on the motionless pin that joins the blades. Relative to your inertial frame, the point of intersection of the blades will soon be moving away from you with a speed greater than that of light. Of course, it is not a material object that is moving, but a geometrical point.

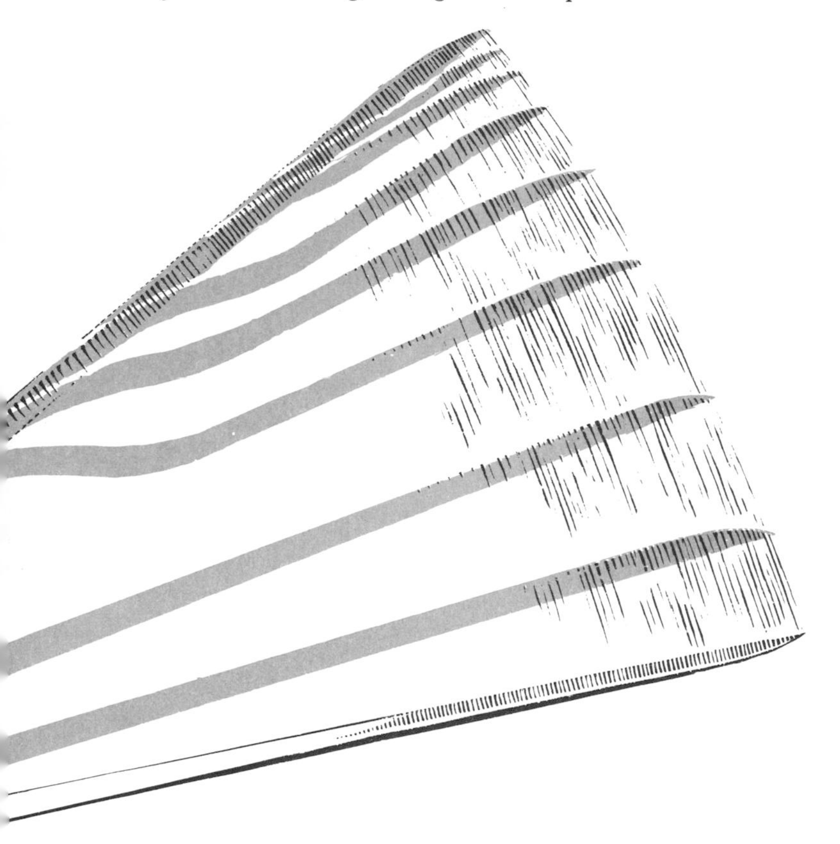

* See Milton A. Rothman, "Things That Go Faster Than Light," *Scientific American* (July 1960).

Perhaps this thought occurs to you: Suppose that the handles of the scissors are on the earth and the point of intersection of the blades is at Neptune. As you wiggle the handles slightly, the intersection point jiggles back and forth. Could you not, then, transmit signals almost instantaneously to Neptune? No, because the impulse that moves the blades has to pass from molecule to molecule, and this transmission must be slower than light. There are no absolutely rigid bodies in special relativity. Otherwise you could simply extend a rigid rod from the earth to Neptune and send messages instantaneously by wiggling one end. There is no way that the giant pair of scissors, or any other type of so-called rigid object, could be used for transmitting a signal with a speed faster than the speed of light.

If a searchlight beam is aimed at a screen that is big enough and far enough away, the searchlight can be turned to make the spot on the screen move across the screen faster than light. Here again, no material object is moving. The motion is really an illusion. If the searchlight is aimed out in space and rotated, distant parts of the beam will sweep through space at a speed far beyond that of light. Oscilloscopes are made in which a beam of light "writes" on the screen with a speed faster than light.

Chapter 5 will show that it is permissible to assume that the earth is a nonrotating frame of reference. From this point of view, the stars will have a circular velocity around the earth that is much greater than the speed of light. A star only ten light-years away has a relative velocity around the earth of twenty thousand times the speed of light. It is not necessary even to look to the stars for this geometrical method of breaking the light barrier. By spinning a top, a child can give the moon a rotational speed (relative to a coordinate

system attached to the top) that far exceeds the velocity of light. Chapter 12 explains that according to a once popular theory about the universe, distant galaxies may be moving away from the earth with a velocity greater than that of light. None of these examples contradicts the assertion that the speed of light is a barrier to sending signals from one material body to another.

An important consequence of the special theory, which can be touched upon only briefly, is that under certain conditions energy will change to mass, and under certain other conditions mass will change to energy. Physicists used to think that the total amount of mass in the cosmos never changes and the total amount of energy never changes. This was expressed by the laws of the "conservation of mass" and the "conservation of energy." Now the two laws have merged into one single law, the "conservation of mass-energy."

When rocket motors accelerate a spaceship, part of the energy goes into the ship's increased relativistic mass. When energy is put into a coffeepot by heating—that is, by speeding up its molecules—the pot actually weighs a trifle more than it did before. As the coffee cools, mass is lost. When a watch is energized by winding, it actually gains a tiny amount of mass. As the watch runs down, it loses the mass. Such gains and losses of mass are so infinitesimal that they would never be considered in the ordinary calculations of physics. The change from mass to energy is not so infinitesimal, however, when a hydrogen bomb explodes!

The bomb's explosion is the sudden conversion to energy of part of the mass of the bomb's material. Energy radiated by the sun has a similar origin. The sun's enormous gravity puts the hydrogen gas in its interior under such great pressure, raises the gas to such a high temperature that hydrogen is fused, or converted, into helium. In this process some mass is turned into energy. The equation that expresses the relation of mass to energy is, as everyone now knows:

$$e = mc^2,$$

where e is energy, m is mass and c^2 is the velocity of light multiplied by itself. This equation was formulated by Einstein in connection with his special theory. It is easy to see from the formula that an exceedingly small bit of mass is capable of releasing a monstrous amount of energy. Life on earth would not exist without the sun's energy, so in a sense life depends on this formula. Now it appears as if the end of life on earth is also bound up with the formula. It is no exaggeration to say that learning how to cope with the terrible fact expressed by this simple equation is the greatest problem that has ever faced mankind.

The bomb, however, is only the most spectacular of many confirmations of the special theory. Experimental evidence began to accumulate almost as soon as the ink was dry on Einstein's 1905 paper. It is, in fact, one of the best-confirmed theories of modern physics. It is confirmed every day in the laboratories of atomic scientists who work with particles that travel with a speed close to that of light. The faster such particles move, the greater the force needed to accelerate them by a given amount: in other words, the greater their relativistic mass. This is precisely why physicists keep building larger and larger machines for accelerating particles. They need stronger and stronger fields to overcome the increasing mass of particles as they are boosted closer and closer to the speed of light. Electrons can now be accelerated to

0.999999999+ the speed of light. This gives to each electron a mass (relative to the earth's inertial frame) that is about forty thousand times its mass at rest! Relativistic changes of time are also observable. The average life of a fast-moving meson, for example, is longer than a slow-moving one because the meson's proper time runs more slowly the faster it goes.

When a particle collides with its antiparticle (a particle of the same structure but opposite electrical charge), there is total and mutual annihilation. The entire mass of both particles turns into radiant energy. So far, this has been done in the laboratory only with individual, short-lived particles. If physicists ever succeed in constructing antimatter (matter made up of antiparticles), they will be able to achieve the ultimate in atomic power. A tiny amount of antimatter on a spaceship, kept suspended by magnetic fields, could be combined slowly with matter to provide propulsion sufficient to carry the ship to the stars.

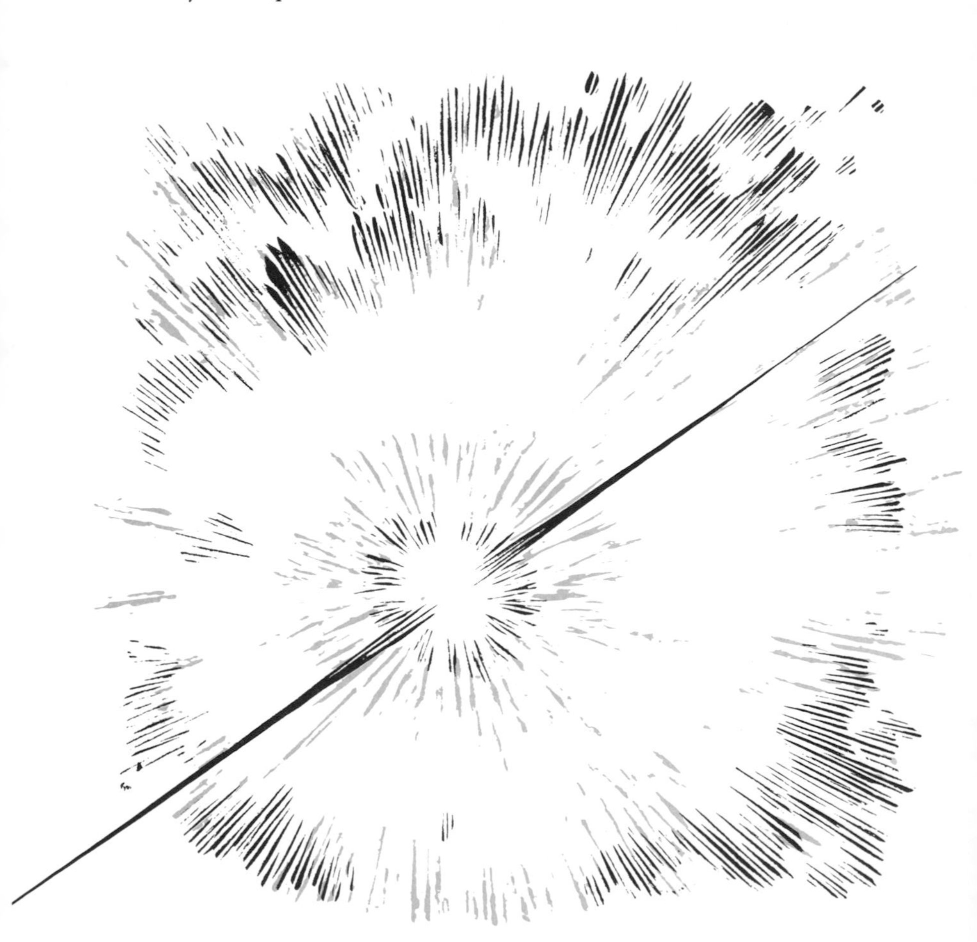

So thoroughly has the special theory of relativity been confirmed by experiment that it would be hard to find a physicist today who doubts the theory's soundness.

Uniform motion is relative. But before it can be said that *all* motion is relative, there is one last hurdle to cross: the hurdle of inertia. Exactly what this hurdle is and how Einstein crossed it will be the topic of Chapter 5.

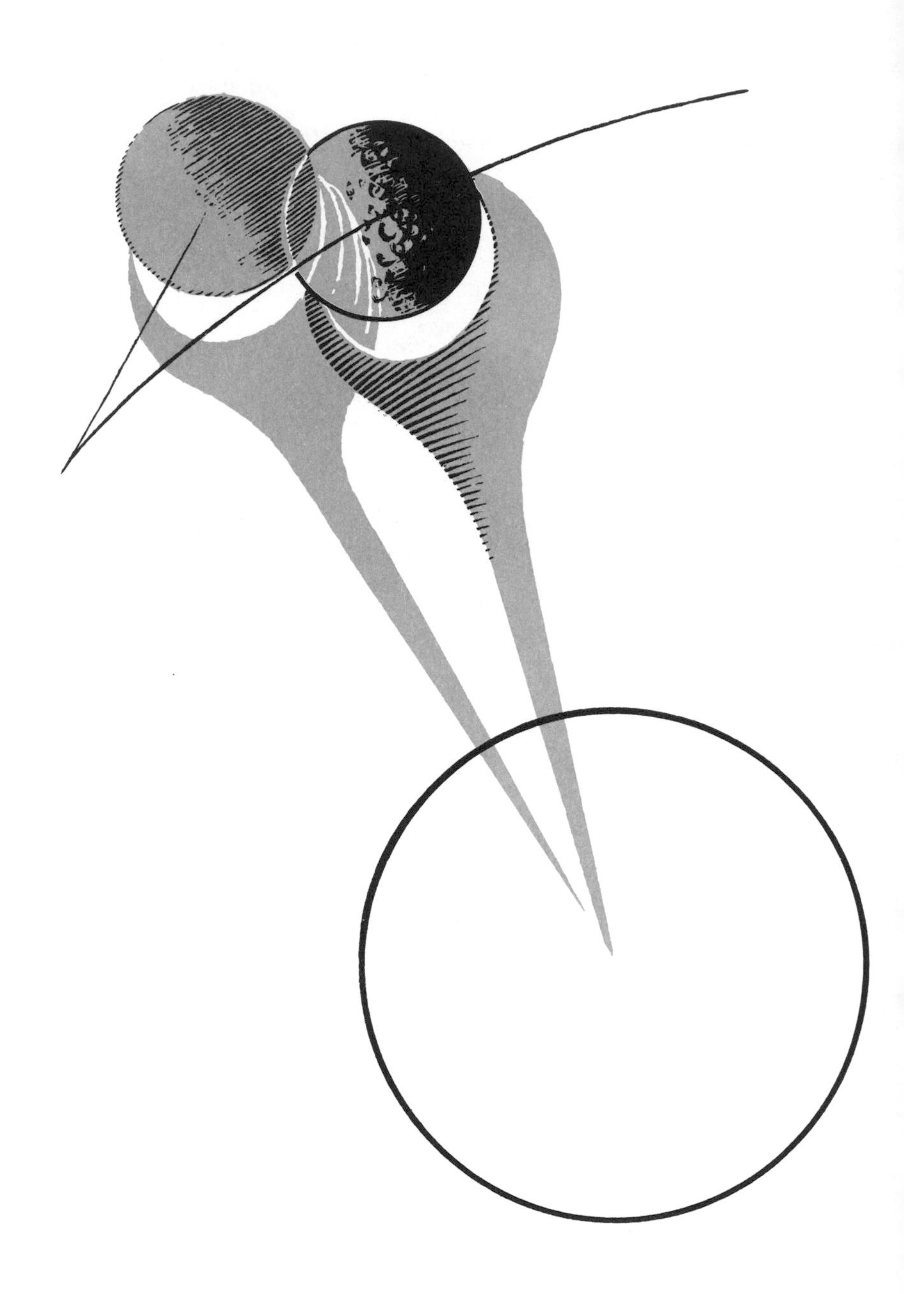

5
The General Theory of Relativity

At the beginning of the second chapter it was pointed out that there are two ways by which absolute motion might be detected: by measuring motion with respect to a beam of light, and by making use of inertial effects that arise when an object is accelerated. The first method was shown by the Michelson-Morley experiment to be unworkable. Einstein's special theory of relativity explained why. This chapter turns to the second method: the use of inertial effects as clues to absolute motion.

When a rocket ship blasts off, an astronaut inside the ship is pressed against the back of his seat with enormous force. This is a familiar inertial effect caused by the rocket's acceleration. Does not this indicate that it is the rocket that is moving? In order to maintain that all motion is relative, including accelerated motion, it must be possible to choose the rocket as a fixed frame of reference. In such a case, the earth and the entire cosmos must be regarded as moving backward, away from the rocket. But if the situation is viewed in this way, how can the inertial forces that act on the astronaut be explained? The force with which he is pressed against his seat seems to indicate, beyond any doubt, that the rocket moves, not the cosmos.

Another convenient example is provided by the rotating earth. Centrifugal force, an inertial effect that accompanies rotation, causes the earth to bulge slightly at the equator. If all motion is relative, does it not follow that the earth can be chosen as a fixed frame of reference, with the cosmos rotating around it? This can certainly be imagined, but then, what would cause the earth's

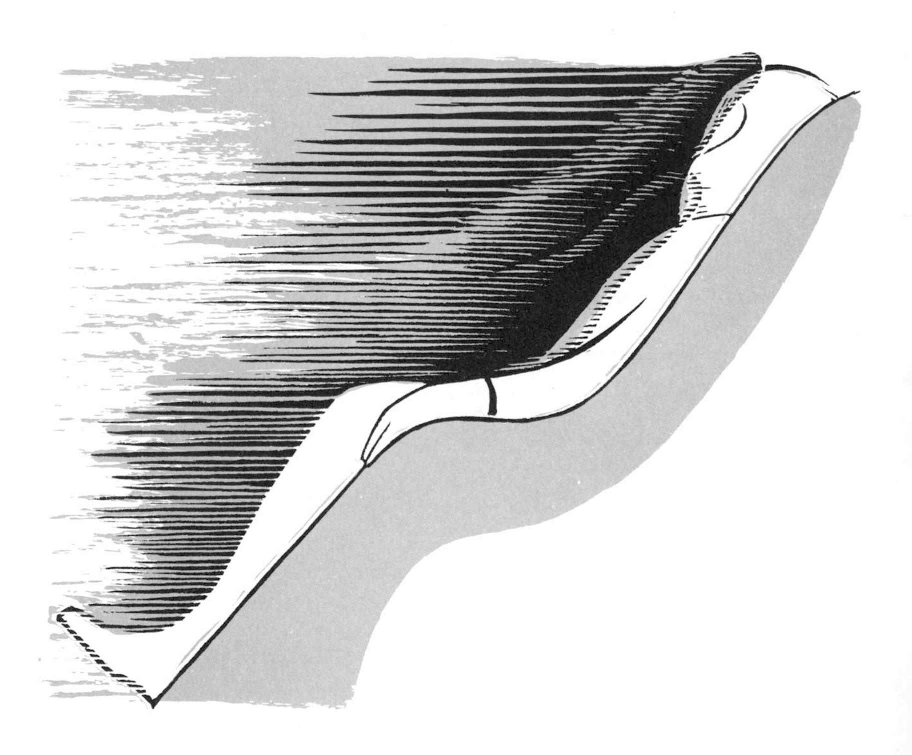

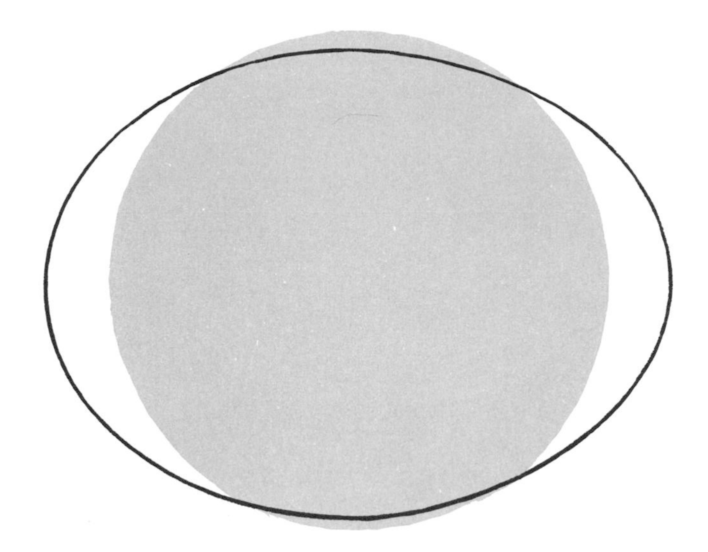

equator to bulge? The bulge seems to indicate that it is the earth, not the universe, that rotates. Incidentally, astronomers are not agreed as to whether centrifugal force continues to maintain the equatorial bulge, or whether the bulge developed in past geologic ages when the earth was more plastic, and has now become a feature of a rigid earth, a feature that would remain even if the earth stopped rotating. All agree, however, that centrifugal force is responsible for the bulge.

This exact line of thought convinced Newton that motion was *not* relative. He cited as proof the fact that if a bucket of water is rotated around a vertical axis, centrifugal force will cause the surface of the water to become concave or even to spill over the sides. It is unimaginable that a rotating universe could have this effect on the water; therefore, it must be concluded, Newton argued, that the bucket's rotation is absolute.

For ten years after he had published his special theory, Einstein brooded about this problem. Most physicists did not even see it as a problem. Why not face the fact, they said, that uniform motion is relative (as the special theory asserts), but that accelerated motion is absolute? Einstein was not satisfied with this state of affairs. He had a hunch that if uniform motion is relative, accelerated motion is also. Finally, in 1916, eleven years after the publication of his special theory, he published his general theory of relativity. The theory is called "general" because it is a generalization or extension of the special theory. It includes the special theory as a special case.

The general theory is a much greater intellectual achievement than the special theory. If Einstein had not been the first to conceive of the special

theory, there is little doubt that other physicists would soon have thought of it. Poincaré, the French mathematician mentioned earlier, was one of several who came within a hair's breadth of it. In a remarkable speech that he gave in 1904,* Poincaré predicted that there would arise "an entirely new mechanics" in which no velocity can exceed that of light, just as no temperature can fall below absolute zero. It would maintain, he said, "the principle of relativity, according to which the laws of physical phenomena should be the same, whether for an observer fixed, or for an observer carried along in a uniform movement of translation; so that we have not any means of discerning whether or not we are carried along in such a motion." Poincaré did not see the essential steps that had to be taken in order to carry out such a program, but he certainly had an intuitive grasp on the essence of the special theory. At the time, Einstein was not aware of how closely the thoughts of Poincaré, Lorentz, and others were to his own. Years later he paid generous tribute to these men.

The general theory of relativity is an altogether different matter. It was, to use Teller's phrase again, "beautifully unexpected": a work of such stupendous originality, along such unorthodox lines, that it came into the scientific world with something like the same effect that the new dance craze, the twist, invaded in 1962 the ballrooms of the United States. Einstein had given a new twist to the ancient dance rhythms of time and space. In a surprisingly short time every physicist in the world was either dancing the new twist, expressing

* This speech was reprinted in *Scientific Monthly* (April 1956).

shocked horror over it, or complaining that he was too old to learn. If Einstein had not lived, no doubt other scientists would have given physics the same twist, but a century or more might have slipped by before they did so. Few other great theories in the history of science seem so completely the work of a single man.

"Newton, forgive me," Einstein wrote toward the end of his life. "You found the only way which, in your age, was just about possible for a man of highest thought and creative power." It is a moving tribute by the greatest scientist of our time to his greatest predecessor.

At the heart of Einstein's general theory is what he calls the principle of equivalence. This is nothing less than the staggering assertion (Newton would have considered it mad) that gravity and inertia are one and the same. This does not mean merely that they have similar effects. *Gravity and inertia are two different words for exactly the same thing.*

Einstein was not the first scientist to be impressed by the strange resemblance between gravitational and inertial effects. Consider for a moment just what happens when a cannonball and a small wooden ball are dropped from the same height. Assume that the cannonball's weight is one hundred times that of the wooden ball. This means that gravity pulls on the cannonball with a force that is one hundred times the force with which it pulls on the wooden ball. It is easy to understand why Galileo's enemies could not believe that two such balls would hit the ground at the same time. Of course we all now know that, ignoring the influence of air resistance, the balls fall side by side. To explain this fact, Newton had to assume something very curious. At the same time that gravity is pulling down on the cannonball, the ball's inertia—that is, its *resistance* to force—is holding back the cannonball. True, the force of gravity is one hundred times greater on the cannonball than on the wooden ball, but the inertia holding back the cannonball is also exactly one hundred times greater!

Physicists often express it this way: The force of gravity on an object is always proportional to the object's inertia. If object A is twice as heavy as object B, its inertia is also twice as great. Twice as much force will be needed to accelerate object A to a certain speed as will be necessary to accelerate object B to the same speed. If this were not the case, objects of different weight would fall with different accelerations.

It is easy to imagine a world in which the two forces are not proportional. In fact, scientists imagined just such a world from the time of Aristotle to the time of Galileo! We could get along quite well in such a world. Conditions would not be exactly the same in a falling elevator, but how often does one ride in a falling elevator? As it is, we happen to live in a world in which the two forces are proportional. Galileo was the first to demonstrate this. Surprisingly accurate experiments confirming Galileo's findings were made around 1900 by a Hungarian physicist named Baron Roland von Eötvös. The most accurate tests of all were made in the early 1960s by Robert H. Dicke and his associates at Princeton University.* As far as they could determine, gravitational mass (weight) is always exactly proportional to inertial mass.

Newton knew, of course, about this curious tug of war between gravity and inertia, a tug of war that causes all objects to fall with the same acceleration, but he had absolutely no way of accounting for it. It was simply an extraordinary coincidence. Because of this coincidence it is possible to make use of inertia in such a way that gravitational fields can be both created and eliminated. Chapter 1 brought out the fact that an artificial gravity field can be produced in a spaceship shaped like a torus (doughnut) simply by rotating the ship like a wheel. Centrifugal force will cause objects inside the ship to press against the outside rim. By rotating the ship at a certain constant speed, an inertial force field is created inside the ship that has the same effect as the gravitational field of the earth. Spacemen would walk about on what they would regard as a curved floor. Dropped objects would fall to this floor. Smoke would rise to the ceiling. All the effects of a normal gravitational field would be present. Einstein illustrated the same point with the following famous thought experiment.

Imagine an elevator that is being pulled up through space with constantly increasing speed. If this acceleration is uniform, and exactly the same as the acceleration with which an object falls to the earth, then persons inside the elevator will believe themselves to be in a gravitational field exactly like the earth's.

Not only can acceleration counterfeit gravity in this way, it can also counteract gravity. In a falling elevator, for example, the downward acceleration completely eliminates the effect of gravity inside the car. A state of zero *g* (zero gravity) prevails inside a spaceship so long as it is in a state of free fall:

* See R. H. Dicke, "The Eötvös Experiment," *Scientific American* (December 1961).

moving freely under the influence of no force except gravity. The weightlessness experienced by Russian and American astronauts on their trips around the earth is explained by the fact that their ships are in a state of free fall as they circle the earth. So long as a spaceship's rocket motors are not working, there is zero *g* inside the ship.

This remarkable correspondence between inertia and gravity remained unexplained until Einstein developed his general theory of relativity. As in his special theory, he invoked the simplest, most daring hypothesis. In special relativity, remember, Einstein said that the reason there seems to be no ether wind is that there *isn't* any ether wind. In general relativity he says: The reason gravity and inertia seem to be the same thing is that they *are* the same thing.

It is not quite correct to say that inside a free-falling elevator a condition of zero gravity is simulated. For an observer on earth it is true that the earth's gravitational field is still there, causing both the elevator and the person inside to fall. But to the observer in the elevator, who takes the elevator as his frame of reference, the earth and the entire universe are accelerating toward him. This sets up a gravity field (as we will see in a moment) that nullifies the field surrounding the earth. The field equations are such that when the total situation is described by the observer in the elevator, the earth's gravitational field has disappeared. It is true zero gravity.

Similarly, it is not quite correct to say that gravity in a rotating spaceship or an upward-accelerating elevator is counterfeited. It is not counterfeited.

Gravity is genuinely created. A gravitational field produced in this way does not have the same geometrical structure as a field surrounding a large body like the earth, but it is a true gravity field nonetheless. As in the special theory, the mathematical description of nature must be made more complicated to permit these startling assertions, but the end result justifies the complication. Instead of two forces, gravity and inertia, there is only one.

Einstein hated complexity and loved simplicity, in his daily life as well as in his thinking. Once when a friend asked him why he refused to buy shaving soap (he shaved with ordinary bar soap) Einstein said he found intolerable the notion of keeping two kinds of soap when one would do. Before Einstein came along with his sharp Occam's razor (the principle that entities should not be multiplied beyond necessity), scientists shaved the universe with two kinds of soap, gravity and inertia. Would Einstein have thought of his general theory had he not found this intolerable?

It may seem strange to use a word like "simplicity" for a theory employing such advanced mathematics that it was once said that no more than twelve men in the world could understand it (an exaggeration, by the way, even at the time the remark was current). The mathematics of relativity is indeed complicated, but this complexity is balanced by a remarkable simplification in the overall picture. The reduction of gravity and inertia to the same phenomenon alone is enough to make general relativity a most efficient way of looking at the world.

Einstein made this point in 1921 when he lectured on relativity at Princeton University. "The possibility," he said, "of explaining the numerical equality of inertia and gravitation by the unity of their nature gives to the general theory of relativity, according to my conviction, such a superiority over the conceptions of classical mechanics, that all the difficulties encountered must be considered as small in comparison."

In addition, relativity theory has what mathematicians like to call "elegance": a kind of artistic grandeur. "Every lover of the beautiful," Lorentz once declared, "must wish it to be true."

Einstein's principle of equivalence—the equivalence of gravity and inertia—makes possible the view that all motion, including accelerated motion, is relative. This is how the trick is done. When Einstein's elevator is visualized as moving upward through the cosmos with accelerating velocity, inertial effects can be observed inside the elevator. But the elevator can theoretically be made a fixed, motionless frame of reference. Now the entire universe, with all its galaxies, is moving down past the elevator with accelerating speed. *This accelerated motion of the universe generates a gravitational field.* The field causes objects inside the elevator to press against the floor. One can say that these effects are gravitational, not inertial.

But which is *really* happening? Is the elevator moving and causing inertial effects or is the universe moving and causing gravitational effects? This is not a proper question. There is no "real," absolute motion. There is only a

relative motion of elevator and universe. This relative motion creates a force field, described by the field equations of the general theory. The field can be called either gravitational or inertial, depending on the choice of a frame of reference. If the elevator is the frame, the field is called gravitational. If the cosmos is the frame, the field is called inertial. Inertia and gravity are merely two different words that can be applied to the same situation. Naturally, it is much simpler, more convenient, to think of the universe as fixed. No one would consider calling the field inside the upward-accelerating elevator gravitational. The general theory of relativity says, however, that the field *can* be called gravitational if a suitable frame of reference is adopted. No experiment that would prove this choice "wrong" can be performed inside such an elevator.

When it is said that the observer in the elevator cannot tell whether the field that is pressing him to the floor is inertial or gravitational, this does *not* mean that he cannot tell the difference between his field and a gravitational field surrounding a large body of matter, such as a planet. The gravitational field around the earth, for example, has a spherical structure that cannot be duplicated by accelerating an elevator in space. If two apples are held a foot apart and dropped from a great height above the earth, they will move closer together as they fall because each apple drops along a straight line aimed toward the center of the earth. In the moving elevator, however, all objects fall along parallel lines.

The difference between the two fields can be brought out by another simple thought experiment. If two apples are dropped in an upward-accelerating elevator, one a meter directly above the other, the distance between them remains constant as they fall. Not so if they are dropped from a height above the earth. The distance between them *lengthens*. It is because the lower object, being closer to the earth's center, is always accelerating faster than the object above it.

Let's combine these two effects and see what happens to a large object of spherical shape as it falls toward the center of a strong gravitational field created by a massive object like the sun. The nonuniformity of the field will squeeze the sphere on the sides and lengthen it in the direction of fall. Astronomers call these "tidal forces." They are the forces that can cause a small planet to disintegrate if it falls toward a much more massive body.

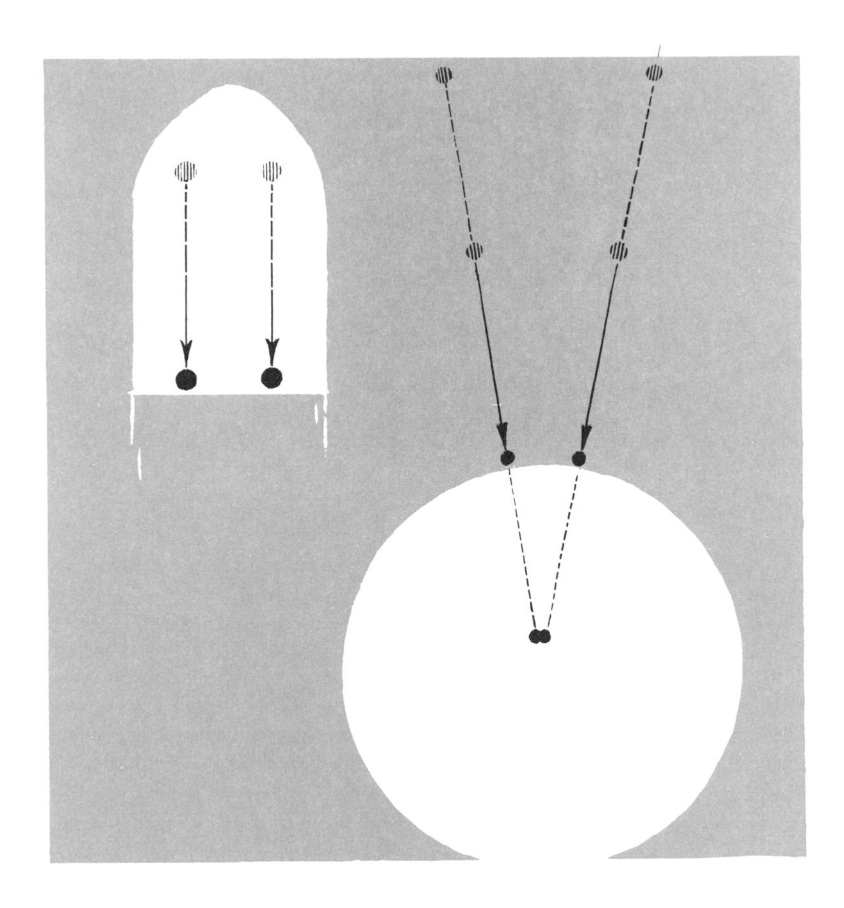

Tidal forces are exerted by gravitational fields surrounding massive bodies, not by gravitational fields produced by accelerations. There is no frame of reference from which an observer would not see the effects of such tidal forces. An observer inside the accelerating elevator would be able to make tests for tidal forces that would tell him the structure of the field. This does not mean that he is distinguishing between inertia and gravity. He is simply distinguishing between fields with different geometrical structures.

A similar situation is presented by the rotating earth. The ancient argument over whether the earth rotates or the heavens revolve around it (as Aristotle taught) is seen to be no more than an argument over the simplest choice of a frame of reference. Obviously, the most convenient choice is the

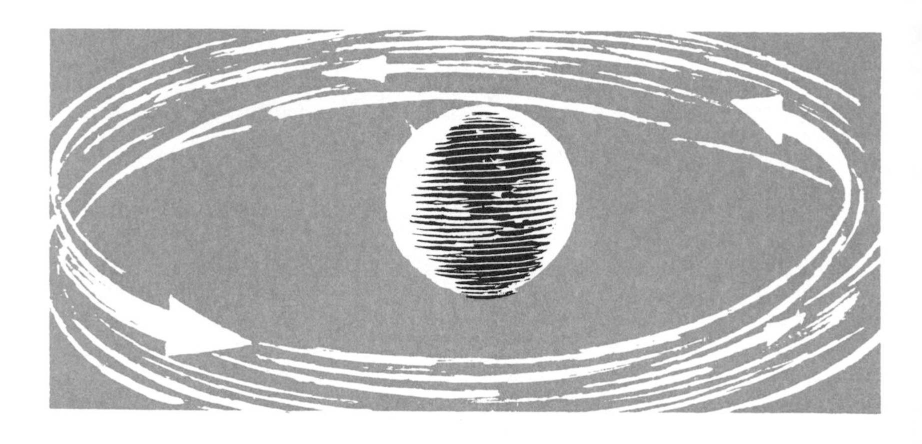

universe. Relative to the universe, we say that the earth rotates and inertia makes its equator bulge. Nothing except inconvenience prevents us from choosing the earth as a fixed frame of reference. In the latter case we say that the cosmos rotates around the earth, generating a gravitational field that acts upon the equator. Again, this field does not have the same mathematical structure as a gravitational field around a planet, but it can be called a true gravitational field nevertheless. If we choose to make the earth our fixed frame of reference, we do not even do violence to everyday speech. We say that the sun rises in the morning, sets in the evening; the Big Dipper revolves around the North Star. Which point of view is "correct"? Do the heavens revolve or does the earth rotate? The question is meaningless. A waitress might just as sensibly ask a customer if he wanted ice cream on top of his pie or the pie placed under his ice cream.

Think of the cosmos as having a kind of mysterious "grip" on every object in it. (Chapter 8 considers the question of where this grip comes from.) The odd thing about this grip is that once an object is moving uniformly through the universe, the universe offers no resistance to the motion. As soon as an attempt is made to force the object into nonuniform (accelerated) motion, the grip tightens. If the universe is made a fixed frame of reference, the grip is called the object's inertia: its resistance to the change of motion. If the object is made a fixed frame of reference, the grip is called gravitational: the universe's attempt to drag the object along as it (the universe) moves in a nonuniform way.

The general theory of relativity is often summed up as follows: Newton made it clear that if an observer is in uniform motion, there is no mechanical experiment he can perform that will prove whether he is moving or at rest. The special theory of relativity extended this to include *all* experiments, optical as well as mechanical. The general theory is another extension: an extension of the special theory to include nonuniform motion. There is no experiment of any sort, the general theory says, by which an observer in *any* sort of motion, uniform or nonuniform, can prove whether he is moving or at rest.

The general theory is sometimes put this way: All the laws of nature are invariant (the same) with respect to any observer. This means that regardless of how an observer is moving, he can describe all the laws of nature (as he sees them) by the same mathematical equations. He may be a scientist working in a laboratory on the earth, or on the moon, or inside a giant spaceship that is slowly accelerating on its way to a distant star. The general theory of relativity provides him with a set of equations by which he can describe all the natural laws involved in any experiment he can perform. These equations will be exactly the same regardless of whether he is at rest, moving uniformly, or moving with acceleration with respect to any other object.

The next chapter takes a closer look at Einstein's theory of gravitation, and how it is related to an important new concept known as spacetime.

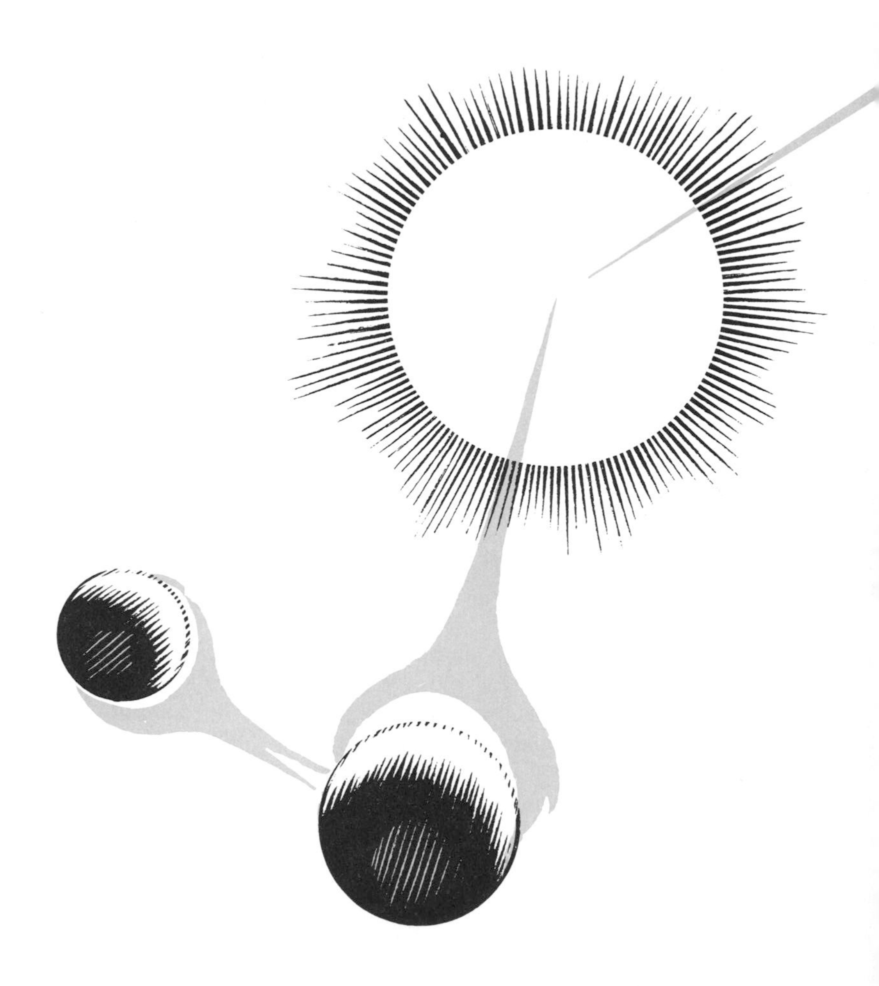

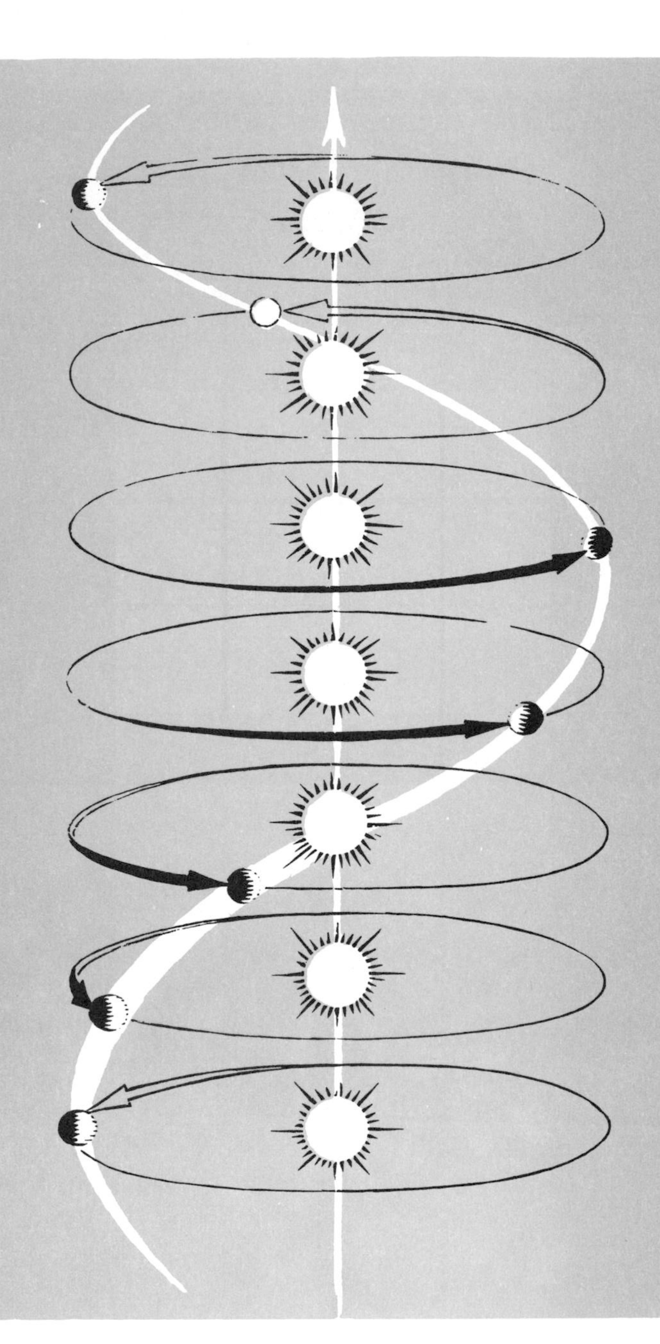

6 Gravity and Spacetime

Before anything can be said about Einstein's theory of gravity, it is necessary to make a few remarks, all too brief, about the fourth-dimension and non-Euclidian geometry. Hermann Minkowski, a Polish mathematician, gave relativity theory its elegant interpretation in terms of a four-dimensional spacetime. Many of the ideas in this chapter are as much Minkowski's as they are Einstein's.

Consider a geometric point. It has no dimension. If it is moved in a straight line, it generates a line of one dimension. Move the line in a direction at right angles to itself and it generates a plane of two dimensions. Move the plane in a direction at right angles to itself and it generates a space of three dimensions. This is as far as we can go in our imagination. But a mathematician can conceive (not in the sense of picturing it in his mind, but in the sense of working out the mathematics) of moving three-dimensional space in a direction at right angles to all three of its dimensions. This generates a Euclidian

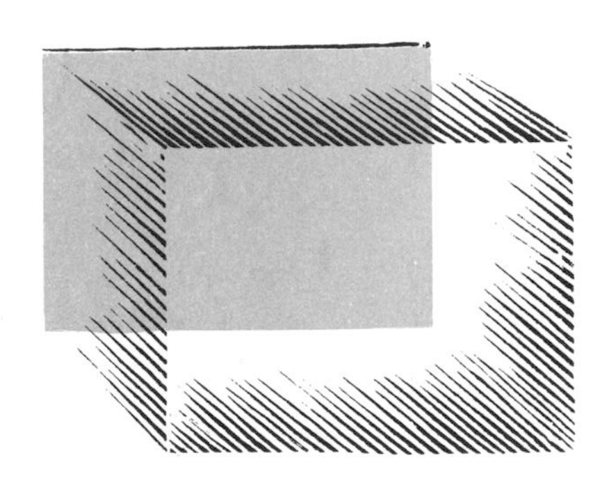

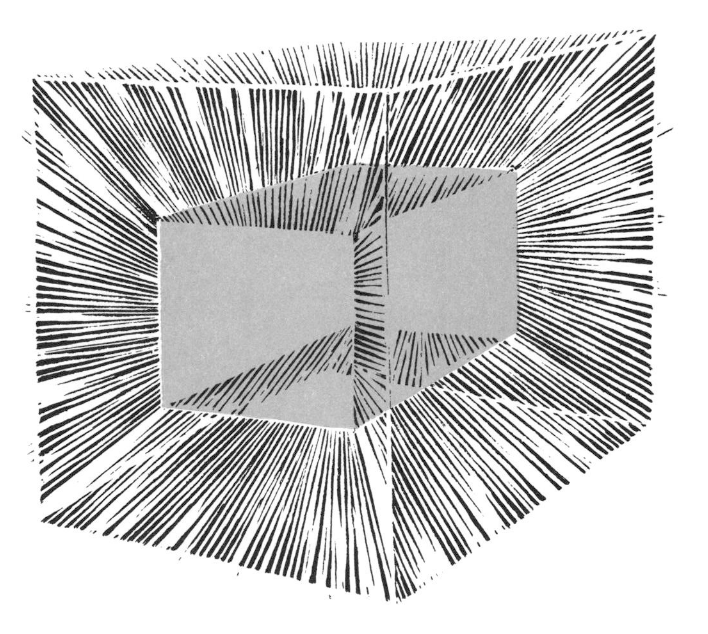

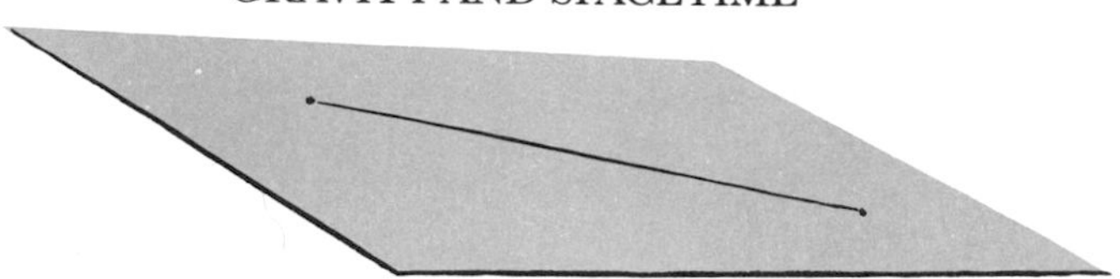

space of four dimensions. There is no need to stop at four. We can go on to spaces of five, six, seven, or more dimensions. All these spaces are Euclidian. They are extensions of Euclidian geometry in the same way that Euclidian solid geometry is an extension of Euclidian plane geometry.

Euclidian geometry is based on a series of postulates of which one is the notorious parallel postulate. This postulate says that on a plane, through a given point outside a line, it is possible to draw one and only one line parallel to the given line. A Euclidian plane to which this postulate applies, is said to be flat. It has zero curvature, infinite area. A non-Euclidian geometry is one in which the parallel postulate is replaced by another postulate. This can be done in two essentially different ways.

One way, called elliptic geometry, says that on the plane *no* line can be drawn through a point outside a line and parallel to that line. The surface of a sphere provides a rough, not exact, model of this type of non-Euclidian plane. The "straightest" possible line on the sphere is a great circle (a circle with a diameter equal to that of the sphere). All great circles intersect each other, so it is impossible for two great circles to be parallel. A non-Euclidian plane of this type is said to have positive curvature. This curvature causes the plane to curve back on itself. It has a finite area instead of an infinite area.

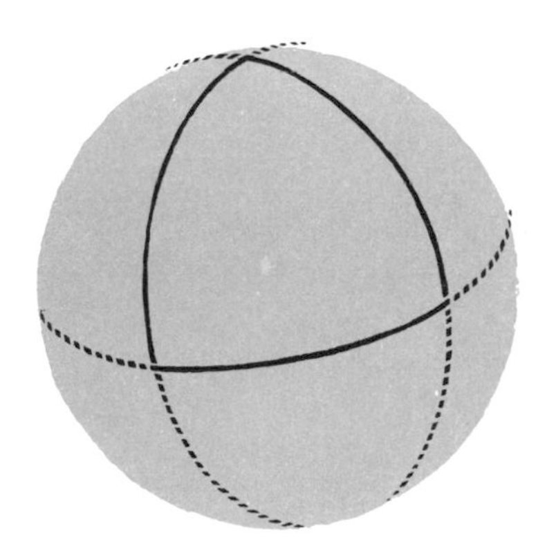

The other type of non-Euclidian geometry, called hyperbolic geometry, is one in which the parallel postulate is replaced by a postulate which says that on a plane, through a point outside a line, there is an infinity of lines that are parallel to the given line. A rough model of a portion of this type of plane is provided by a saddle-shaped surface. Such a surface is said to have negative curvature. It does not close back on itself. Like the Euclidian plane, it extends to infinity in all directions.

Both elliptic and hyperbolic geometry are non-Euclidian geometries of constant curvature. This means that the curvature is everywhere the same; objects do not undergo distortions as they move from one spot to another. A more general type of non-Euclidian geometry, usually called *general Riemannian geometry*, is one that permits the curvature to vary from point to point in any specified way.

Just as there are Euclidian geometries of 2, 3, 4, 5, 6, 7, . . . , dimensions, so also there are non-Euclidian geometries of 2, 3, 4, 5, 6, 7, . . . , dimensions.

In developing the general theory of relativity, Einstein found it necessary to adopt a four-dimensional general Riemannian geometry. Instead of a fourth space dimension, however, Einstein made *time* his fourth dimension.

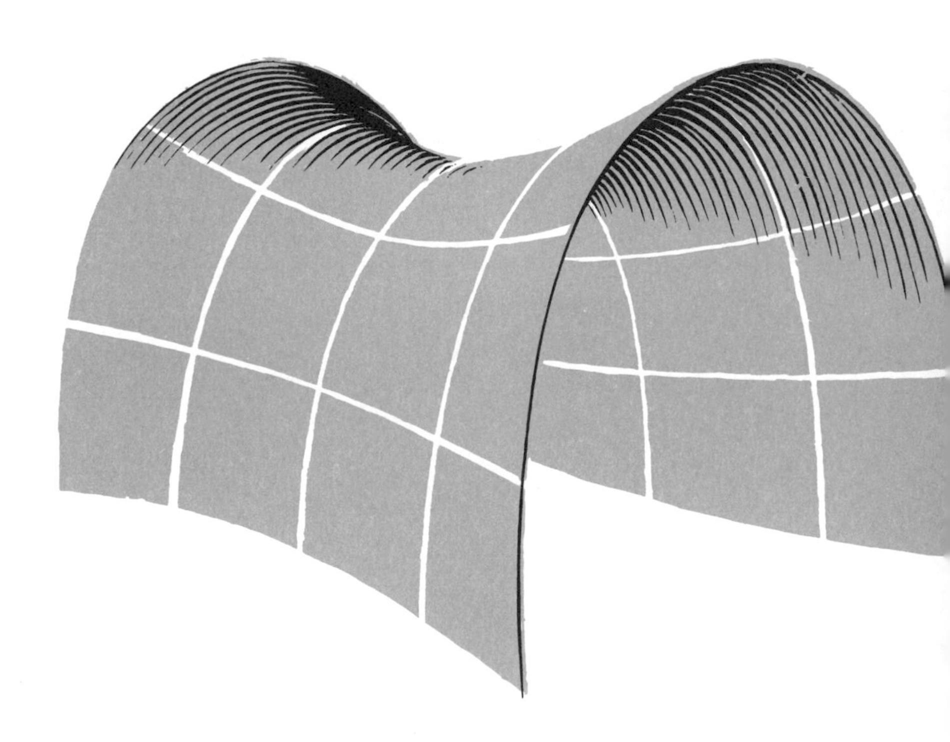

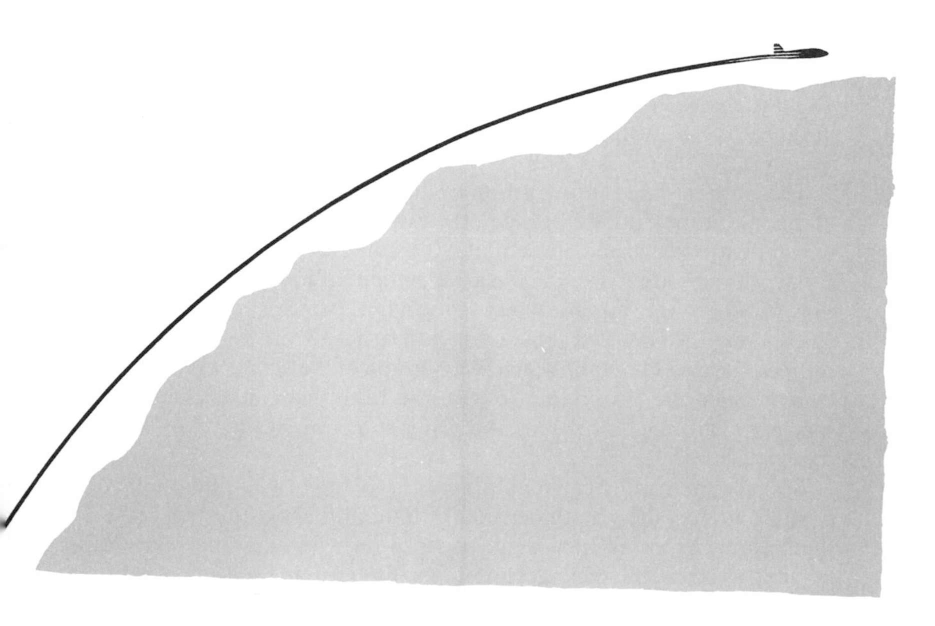

There is nothing mysterious or occult about this concept. It merely means that every event that takes place in the universe is an event occurring in a four-dimensional world of spacetime.

This can be made clear by considering the following event. You get into a car at 2 P.M. and drive from your home to a restaurant that is 3 kilometers south and 4 kilometers east of your house. On the two-dimensional plane the actual distance from your house to the restaurant is the hypotenuse of a right triangle with sides of 3 and 4 kilometers. This hypotenuse has a length of 5 kilometers. But it also took you a certain length of time, say, ten minutes, to make this drive. This time span can be shown on a three-dimensional graph.

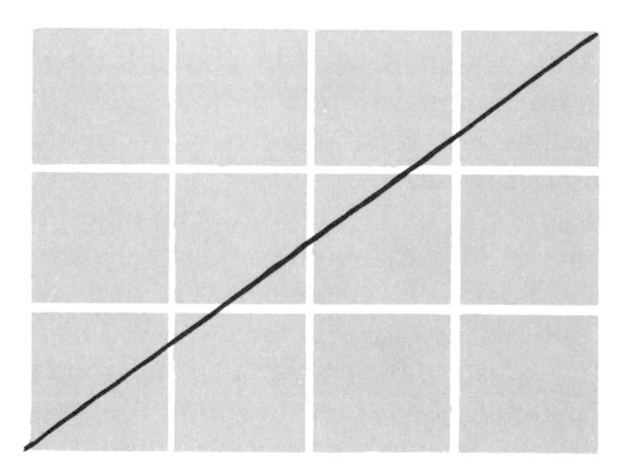

One coordinate of the graph is the distance south in kilometers, another is the distance east in kilometers; the vertical coordinate is the time in minutes. On this three-dimensional graph of spacetime, the "interval" (spacetime distance) between the two events (leaving your house and arriving at the restaurant) is shown as a straight line.

This straight line is not a graph of the actual trip. It is simply a measure of the spacetime distance between the two events. A graph of the actual trip would be a complicated curved line. It would be complicated because your car accelerates when it starts, the arrangement of streets may make it impossible to drive to the restaurant on a straight line, perhaps you stopped at traffic lights along the way, and finally, you had to accelerate negatively when you stopped the car. The complicated wavy graph of the actual trip is, in relativity theory, called the "world line" of the trip. In this case, it is a world line in a spacetime of three dimensions, or (as it is sometimes called) in a Minkowski three-space.

Because the trip by car took place on a plane of two dimensions, it was possible to add the one dimension of time and show the trip on a three-dimensional graph. When events occur in three-dimensional space it is not possible to draw an actual graph of four-dimensional spacetime, but mathematicians have ways of handling such graphs without actually drawing them. Try to imagine a four-dimensional hyperscientist who can construct four-dimensional graphs as easily as the ordinary scientist can construct graphs with two and three dimensions. Three of the coordinates of his graph are the three dimensions of our space. The fourth coordinate is our time. If a spaceship leaves the earth and lands on Mars, our imaginary hyperscientist will draw the world line of this trip as a curve on his four-dimensional graph. (The line is curved because the ship cannot make such a trip without accelerating.) The spacetime "interval" between takeoff and landing will appear as a straight line on the graph.

In relativity theory, every object is a four-dimensional structure lying timelessly along its world line in the four-dimensional world of spacetime. If an object is considered at rest with respect to the three space coordinates, it is still traveling through the dimension of time. Its world line will be a straight line that is parallel with the time axis of the graph. If the object moves through space with uniform motion, its world line will still be straight, but no longer parallel with the time axis. If the object moves with nonuniform motion, its world line becomes curved.

Strictly speaking, one should not say that an object moves along its world line, because "moves" implies movement in time, whereas time is *already* represented by the world line. The world line is no more than a convenient way to graph the motions of an object in three-space. The fact that a Minkowski graph is, in a sense, a static, timeless picture of the world has nothing whatever to do with the question of whether the future is or is not completely determined by the present. An object moving in a random, unpredictable way can

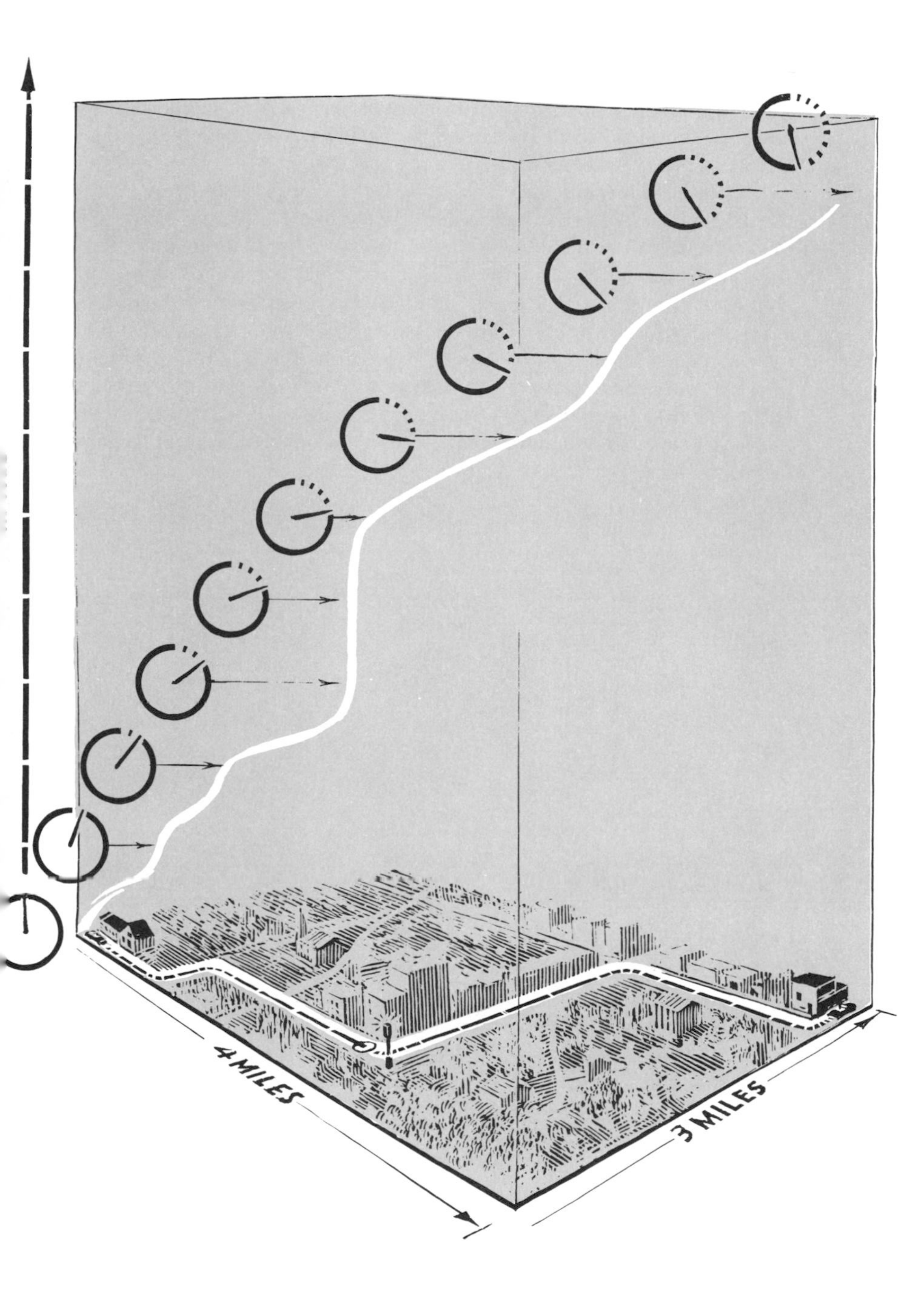
4 MILES
3 MILES

be graphed by a world line just as easily as an object moving in a predictable way. After an event has occurred, its Minkowski graph does indeed freeze the event in a timeless "block universe," but this has no bearing on the question of whether the event had to happen the way it did.

We are now in position to look at the Lorentz-FitzGerald contractions of the special theory from a new point of view: the Minkowski point of view, or the viewpoint of our hyperscientist. As we have seen, when two spaceships pass each other in relative motion, observers on each ship see certain changes in the shape of the other ship as well as changes in the rate of the other ship's clock. This is because space and time are not absolutes that exist independently of each other. They are, so to speak, like shadow projections of a four-dimensional spacetime object. If a book is held in front of a light and its shadow projected on a two-dimensional wall, a turn of the book will alter the shape of its shadow. With the book in one position the shadow is a fat rectangle.

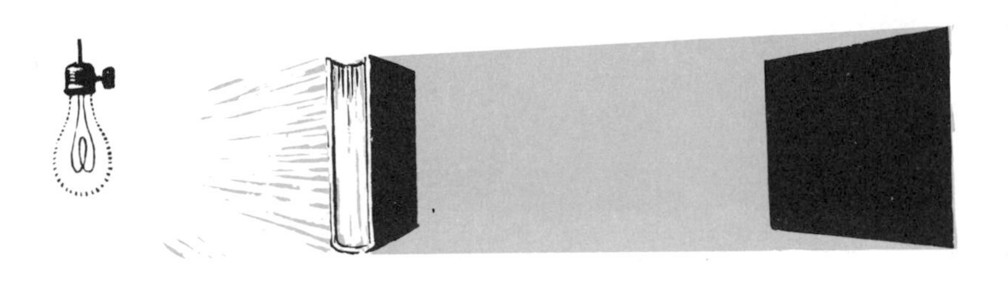

In another position it is a thin rectangle.

The book does not change its shape; only its two-dimensional shadow changes. In a similar way, an observer sees a four-dimensional structure, say, a spaceship, in different three-dimensional projections depending on his motion relative to the structure. In some cases, the projection shows more of space and less of time; in other cases, the reverse is true. The changes that he observes in the space and time dimensions of the other ship can be explained by a kind of "rotation" of the ship in spacetime, causing its shadow projections in space and time to alter. This is what Minkowski had in mind when (in 1908) he began a famous lecture to the 80th Assembly of German Natural Scientists and Physicians. This lecture is reprinted in *The Principle of Relativity*, by Albert Einstein and others. No popular book on relativity is complete without this quotation:

> The views of space and time which I wish to lay before you have sprung from the soil of experimental physics, and therein lies their strength. They are radical. Henceforth space by itself, and time by itself, are doomed to fade away into mere shadows, and only a kind of union of the two will preserve an independent reality.

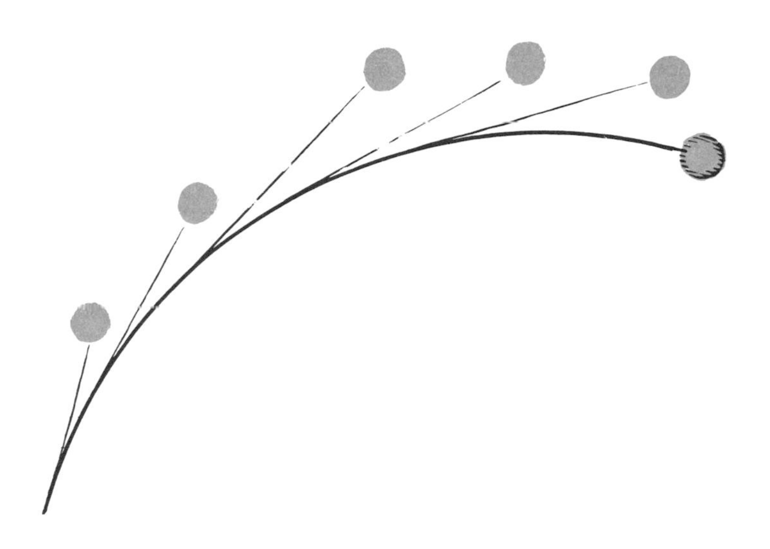

The important point to grasp here is that the spacetime structure, the four-dimensional structure, of the spaceship is just as rigid and unchanging as it is in classical physics. This is the essential difference between the discarded Lorentz contraction theory and the Einstein contraction theory. For Lorentz, the contraction was a real contraction of a three-dimensional object. For Einstein, the "real" object is a four-dimensional object that does not change at all. It is simply seen, so to speak, from different angles. Its three-dimensional projection in space and its one-dimensional projection in time may change, but the four-dimensional ship of spacetime remains rigid.

Here is another instance of how the theory of relativity introduces new absolutes. The four-dimensional shape of a rigid body is an absolute, unchanging shape. We can slice spacetime so the shape of a spaceship depends on the motion of the frame of reference from which we make the slice, but (as J. J. C. Smart writes in the introduction to his anthology, *Problems of Space and Time*), "the fact that we can take slices at different angles through a sausage does not force us to give up an absolute theory of sausages." The theory of relativity, Smart continues, does not decide between absolute and relational philosophies of space and time. It merely shifts the question from space and time taken separately to an interrelated spacetime continuum.

In a similar sense the four-dimensional interval between any two events in spacetime is an absolute interval. Observers moving at great speeds and with different relative motions may disagree on how far apart they judge two events to be in space, and on how far apart they judge two events to be in time, but *all* observers, regardless of their motions, will agree on how far apart they judge two events to be in spacetime. E. F. Taylor and J. A. Wheeler, in their marvelous textbook *Spacetime Physics*, put it this way: "Space is different for different observers. Time is different for different observers. Spacetime is the same for everyone."

In classical physics an object moves through space in a straight line, with uniform velocity, unless acted upon by a force. A planet, for example, would move off in a straight line were it not held by the force of the sun's gravity. From this point of view, the sun is said to "pull" the planet into an elliptical orbit.

In relativity physics an object also moves in a straight line, with uniform velocity, unless acted upon by a force, but the straight line must be thought of as a line in spacetime instead of space. This is true even in the presence of gravity. The reason for this is that gravity, according to Einstein, is not a force at all! The sun does not "pull" on the planets. The earth does not "pull" down the falling apple. What happens is that a large body of matter, such as the sun, causes spacetime to curve in the area surrounding it. The closer to the sun, the greater the curvature. In other words, the structure of spacetime in the neighborhood of large bodies of matter becomes non-Euclidian. In this non-Euclidian space, objects continue to take the straightest possible paths, but what is straight in spacetime is seen as curved when projected onto space. Our imaginary hyperscientist, if he plots the orbit of the earth on his four-dimensional graph, will plot it as a "straight" line. We who are three-dimensional creatures (more precisely, creatures who split up spacetime into

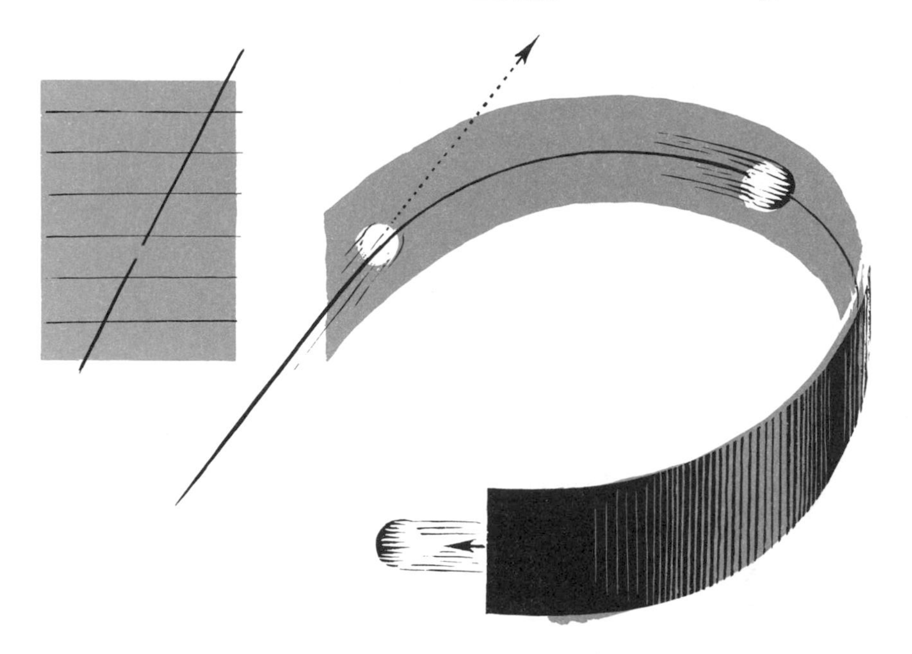

three-dimensional space and one-dimensional time) see the space path as an ellipse.

Writers on relativity theory often explain it in the following way. Imagine a rubber sheet stretched out flat like a trampoline. A grapefruit placed on this sheet will make a depression. A marble placed near the grapefruit will roll toward it. The grapefruit is not "pulling" the marble. Rather, it has created a field (the depression) of such a structure that the marble, taking the path of least resistance, rolls toward the grapefruit. In a roughly (very roughly) similar way, spacetime is curved or warped by the presence of large masses like the sun. This warping is the gravitational field. A planet moving around the sun is not moving in an ellipse because the sun pulls on it, but because the field is such that the ellipse is the "straightest" possible path the planet can take in spacetime.

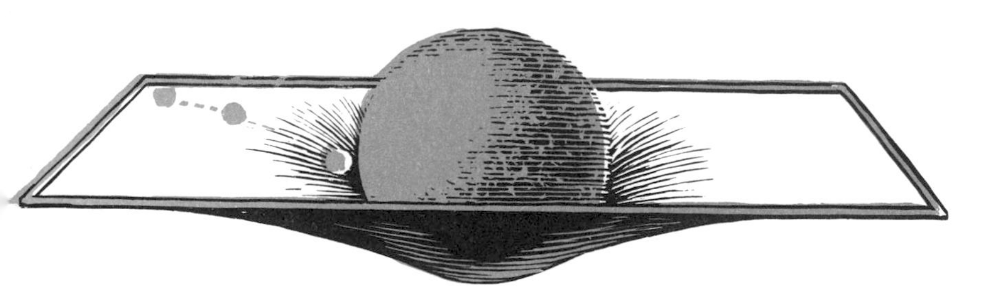

Such a path is called a geodesic. This is such an important word in relativity theory that it should be explained more fully. On a Euclidian plane, such as a flat sheet of paper, the straightest distance between two points is a straight line. It is also the shortest distance. On the surface of a globe, a geodesic between two points is the arc of a great circle. If a string is stretched as tautly as possible from point to point, it will mark out the geodesic. This, too, is both the "straightest" and the shortest distance connecting the two points.

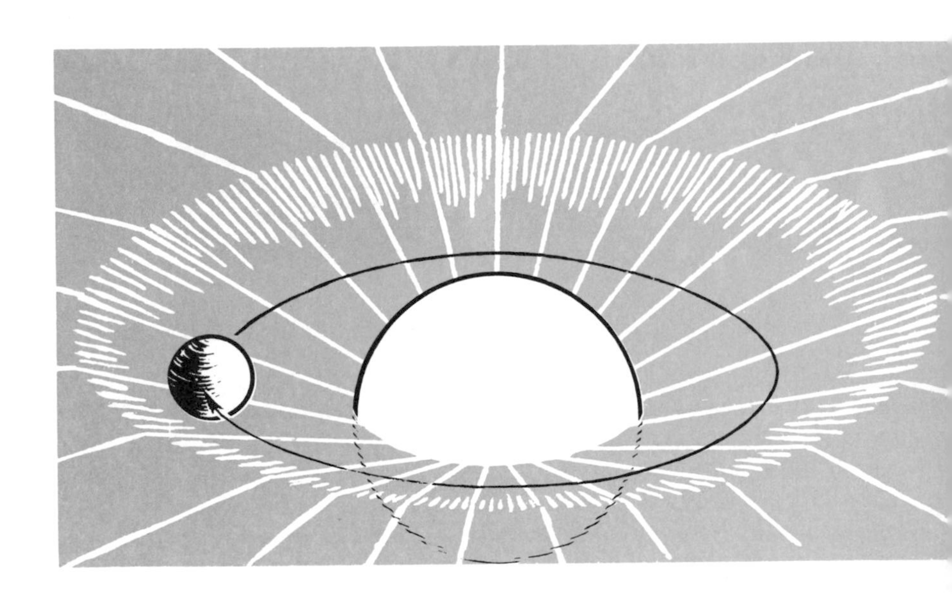

In a four-dimensional *Euclidian* geometry, where all the dimensions are space dimensions, a geodesic also is the shortest and straightest line between two points. But in Einstein's *non-Euclidian* geometry of spacetime, it is not so simple. There are three space dimensions and one time dimension, united in a way that is specified by the equations of relativity. This structure is such that a geodesic, although still the straightest possible path in spacetime, is the *longest* instead of the shortest distance. This concept is impossible to explain without going into complicated mathematics, but it has this curious result: A body moving under the influence of gravity alone always finds the path along which it takes the longest proper time to travel; that is, the longest when measured by its own clock. Bertrand Russell has called this the "law of cosmic laziness." The apple falls straight down, the missile moves in a parabola, the earth moves in an ellipse because they are too lazy to take other routes.

It is this law of cosmic laziness that causes objects to move through space in ways sometimes attributed to inertia, sometimes to gravity. If you tie a string to an apple and swing it in circles, the string keeps the apple from moving in a straight line. We say that the apple's inertia pulls on the string. If the string breaks, the apple takes off in a straight line. Something like this happens when an apple falls off a tree. Before it falls, the branch prevents it from moving

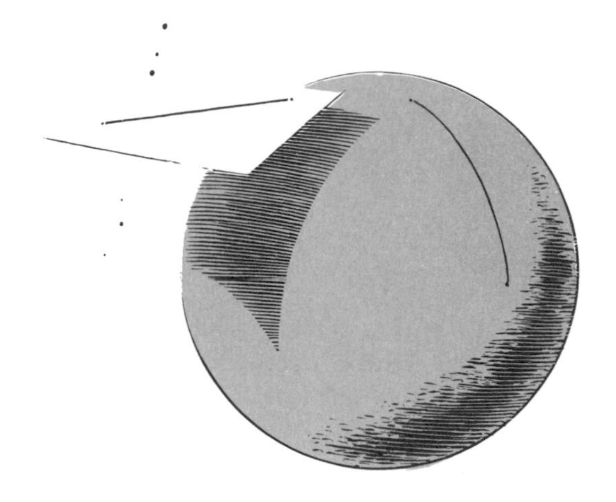

through space. The apple on the branch is at rest (relative to the earth) but speeding along its time coordinate because it is constantly getting older. If there were no gravitational field, this travel along the time coordinate would be graphed as a straight line on a four-dimensional graph. But the earth's gravity is curving spacetime in the neighborhood of the apple. This forces the apple's world line to become a curve. When the apple breaks away from the branch, it continues to move through spacetime, but (being a lazy apple) it now "straightens" its path and takes a geodesic. We see this geodesic as the apple's fall and attribute the fall to gravity. If we like, however, we can say that the apple's inertia, after the apple is suddenly released from its curved path, carries it to the ground.

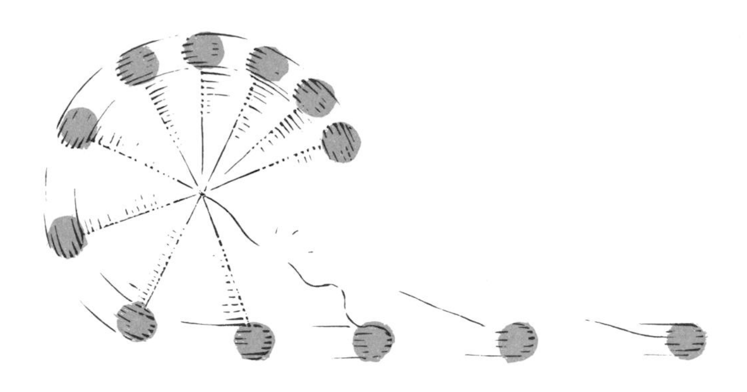

After the apple falls, suppose a boy comes along and kicks it with his bare foot. He shouts in pain because the kick hurts his toes. A Newtonian would say that the apple's inertia resisted his kick. An Einsteinian can say the same thing, but he can also say, if he prefers, that the boy's toes caused the entire cosmos (including the toes) to accelerate backward, setting up a gravitational field that pulled the apple with great force against his toes. It is all a matter of words. Mathematically the situation is described by one set of spacetime field equations, but it can be talked about informally (thanks to the principle of equivalence) in either of two sets of Newtonian phrases.

Although relativity theory replaces gravity by a geometrical warping of spacetime, it leaves many basic questions unanswered. Does this warping take place instantaneously through space or does it propagate like a wave motion? Almost all physicists agree that the warping moves like a wave and that these waves travel with the speed of light. There is also good reason to believe that gravity waves consist of tiny indivisible particles of energy called "gravitons." In 1969 Joseph Weber, at the University of Maryland, announced that his equipment, consisting of huge aluminum cylinders, had detected gravity radiation. It seemed to be coming from cataclysmic events at the center of the Milky Way. Since then, dozens of attempts have been made to confirm Weber's claim, some by physicists with detecting equipment more sensitive than Weber's. The results have been negative. The present consensus is that Weber misinterpreted his readings, and that gravity waves have not yet been observed.

As for gravitons, no one has any knowledge of what a graviton is like, although many physicists are trying to invent theories that will predict some of its properties. Presumably it contains a tiny bit of spacetime curvature, otherwise large numbers of gravitons would be unable to transmit curvature through space. At the moment the graviton, like the particle physicists' quark, remains a hypothetical beast that physicists hope someday to capture.

In 1939 the famous British mathematician and physicist P. A. M. Dirac developed a theory in which gravity is slowly weakening as the universe expands and its density thins. Indeed, the weakening of gravity could even be the cause, or partial cause, of the universe's expansion. (We will discuss this expansion in later chapters.) Many physicists and astronomers take this notion seriously and have constructed similar theories.* If gravity is weakening, large bodies in the universe would tend to expand. This could explain the cracking of the crusts of the moon, and planets like the earth and Mars, and contribute to the drifting of continents on the earth. The sun, also, would be expanding. Two billion years ago it would have been smaller, denser, and hotter—a fact that would explain the tropical conditions that seem to have prevailed over most of the earth in earlier geological epochs.

Relativity theory furnishes a new way of looking at gravity and describing it, but it still remains a mysterious, little-understood phenomenon. No one knows what connection it has, if any, with electromagnetism. Einstein and others have tried to develop a "unified field theory" that will unite gravity and electromagnetism in one set of mathematical equations. The results were disappointing until the 1980s, when a variety of "theories of everything" were proposed. Many physicists, including Stephen Hawking, believe that a final unification of all the particles and forces of nature is close at hand.

* See Thomas C. Van Flandern, "Is Gravity Getting Weaker?," *Scientific American* (February 1976), and "Is the Gravitational Constant Constant?," Chapter 9 in Clifford Will's *Was Einstein Right?* (revised edition, Basic Books, 1993.)

7

Tests of General Relativity

Has the general theory of relativity been supported by experimental evidence? When the first edition of this book was published, physicists were complaining about the weakness of such evidence in contrast to the strong, abundant evidence for the special theory. As Misner, Thorne, and Wheeler put it in their big book *Gravitation:* "For the first half-century of its life, general relativity was a theorist's paradise, but an experimentalist's hell. No theory was thought more beautiful, and none was more difficult to test."

This is no longer true. Advances in technology, especially in the precision of instruments for measuring time, have made it possible to test general relativity in a variety of ways, and it is safe to predict that more and better tests will be made before the nineties run their course. What follows is a brief sketch of how Einstein's theory of gravitation, the heart of the general theory, has remained undamaged by experimental evidence over the past sixty years.

The first great test of general relativity involved the rotation of Mercury's elliptical orbit. Mercury's orbit departs more from a circle than the orbit of any other planet, with the sole exception of Pluto. This makes it much easier to measure the orbit's slow rotation, as predicted by both the Newtonian and Einsteinian theories of gravity. The major axis of Mercury's orbit wheels around the sun at a rate close to 5,600 seconds of arc per century. Newton's equations for gravity, after taking into account the influence of other planets, lead to an expected rotation of about 43 seconds per century *less* than what is actually observed. Einstein's equations give the tiny planet an additional relativistic push, so to speak, of just the right amount—43 seconds of arc per century.

Is this a dramatic confirmation of general relativity? Most physicists think it is, but Dicke, the Princeton physicist mentioned in Chapter 5, is not so sure. He and his former student, Carl Brans, have a theory of gravity in which Einstein's tensor field, although it accounts for about 95 percent of gravity, is combined with a scalar field similar to Newton's. Dicke's theory, known as a scalar-tensor theory, leads to a prediction that the relativistic push on the axis of Mercury's orbit should be less than the 43 seconds supplied by Einstein's theory. The Brans-Dicke theory is the leading contender among new theories of gravitation that modify general relativity in significant ways.

Is it possible, Dicke asked himself, that something has been overlooked in measuring the various gravitational fields acting on Mercury; something that would explain the discrepancy between the prediction of his theory and Mercury's actual orbit? At one time, before Einstein's theory accounted so beautifully for the excess rotation of Mercury's orbit, some astronomers conjectured that a tiny planet—it was even given the name of Vulcan—is hugging the sun inside Mercury's orbit and giving Mercury the needed push. This was soon ruled out by modern telescopes capable of seeing such a planet. Other possible sources for the push were suggested and discredited. Only one good possibility remains: Perhaps the sun is not perfectly round but, like the earth, flattened at the poles. The bulge would increase the sun's gravitational influence on Mercury, making the relativistic effect on the planet's orbit closer to what Dicke's scalar-tensor theory says it should be.

In 1964 Dicke and his associates began work on a device for measuring the sun's shape. It consisted essentially of a wheel with two notches on opposite sides. The plan was to spin this wheel in front of the sun's image so that it would screen off all light except light coming from the rim. A bulging equator would send more light through the rotating notches than would the flattened poles, and this variation in brightness would cause a flickering. By measuring the flicker, the shape of the sun could be calculated with higher precision, Dicke believed, than ever before.

Dicke announced his results in 1967. The sun, he said, is indeed oblate, and to a degree surprisingly close to what his theory predicted. This oblateness, he reasoned, could be caused by a rapidly rotating inner core. Strong magnetic fields on the sun's surface, interacting with the gas that envelops the sun, could exert a braking effect that would slow the surface rotation to its observed rate of about once every twenty-eight days.

Unfortunately, many anomalies have turned up in later attempts to measure the sun's shape, and these anomalies cast grave doubts on the accuracy of Dicke's figures. Dicke assumed that greater brightness at the sun's equator is a measure of oblateness. Astronomers now believe that the brightness has other causes. In 1974 Henry Allen Hill, at the University of Arizona, reported on observations which led him to conclude that the increased brightness of the sun's equator is just that, and has nothing to do with the sun's shape. A year later he reported that the distribution of brightness on the sun's disk oscillates at various frequencies in periods of a few minutes to an hour. More recent observations have indicated that the sun is indeed pulsing like an enormous heart, but at the moment astronomers are not agreed on the nature of the pulses or their causes. At any rate, Dicke's effort to discredit Einstein's gravitational theory by finding a fatter sun is now regarded as unsuccessful.

A second major prediction made by general relativity was that light from the sun ought to show an extremely minute shift toward the red portion of its spectrum. According to Einstein's equations, strong gravitational fields have a slowing effect on time. This means that any rhythmic process, such as the vibrations of atoms or the ticking of a balance-wheel clock, would take place on the sun at a slightly lower rate than on the earth. This would shift the spectrum of sunlight toward the red. Such a shift was observed, but it was not until 1962 that it was measured with sufficient accuracy to provide good confirmation for the general theory. A white dwarf star very close to Sirius (the Dog Star), known as the companion of Sirius, is much denser than our sun, and so should have a greater redshift. Sir Arthur Stanley Eddington was elated when such a shift was observed, but the data later proved to be unreliable. Recent measurements of the redshifts of white dwarfs have been more successful. In the early 1960s, using the Mössbauer effect (see Chapter 9), a redshift of gamma rays, in a 22.5-meter-high tube at Harvard University, agreed to a precision of 1 percent with general relativity's prediction. Other recent tests of gravity's effect on time will be discussed in Chapter 9.

The most dramatic of all early tests of the general theory took place in 1919 during a total eclipse of the sun. Einstein had reasoned as follows: if an elevator in interstellar space were pulled upward with an accelerating velocity, a light beam traveling from side to side inside the elevator would bend down in a parabolic path. This would be regarded as an inertial effect, but

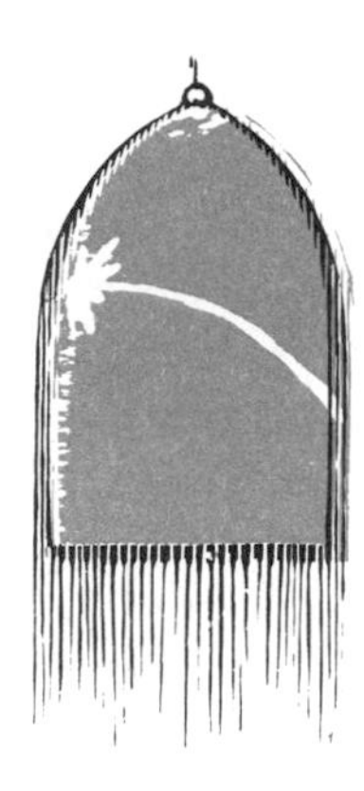

according to the general theory, one can make the elevator a fixed frame of reference and view the curving of the beam as a gravitational effect. Gravity, then, is capable of curving light beams. The curving is much too minute to be detected by any laboratory experiment, but it can be measured by astronomers during a total eclipse of the sun. Because the sun's light is blocked off by the moon, stars very close to the sun's edge become visible. Light from these stars passes through the strongest part of the sun's gravitational field. Any shift in the apparent positions of these stars would indicate that the sun's gravity was bending their light. The greater the shift, the greater the bend.

A word of caution: when you read about the "bending" of light by gravity or inertia, you must remember that this is just a three-dimensional way of speaking. In space the light does indeed curve. But in four-dimensional spacetime, light continues, as in classical physics, to move along geodesics. It takes the "straightest" possible path.

Eddington, the English astronomer, was in charge of an expedition of scientists that went to Africa in 1919 to observe the total eclipse of the sun. The primary purpose of the expedition was to make accurate measurements of the positions of stars close to the sun's rim. Newton's physics also suggested a bending of light in gravitational fields, but Einstein's equations predicted a deflection about twice as large. So there were at least three possible outcomes of the test:

1. There would be no change in the positions of the stars.
2. The deflection would be close to what Newtonian physics had predicted.
3. The deflection would be close to what Einstein had predicted.

The first outcome would damage both Newton's equations and those of the general theory of relativity. The second would strengthen Newton, discredit Einstein. The third would discredit Newton, strengthen Einstein. According to a story that made the rounds at the time, two astronomers on the expedition were discussing the three possibilities.

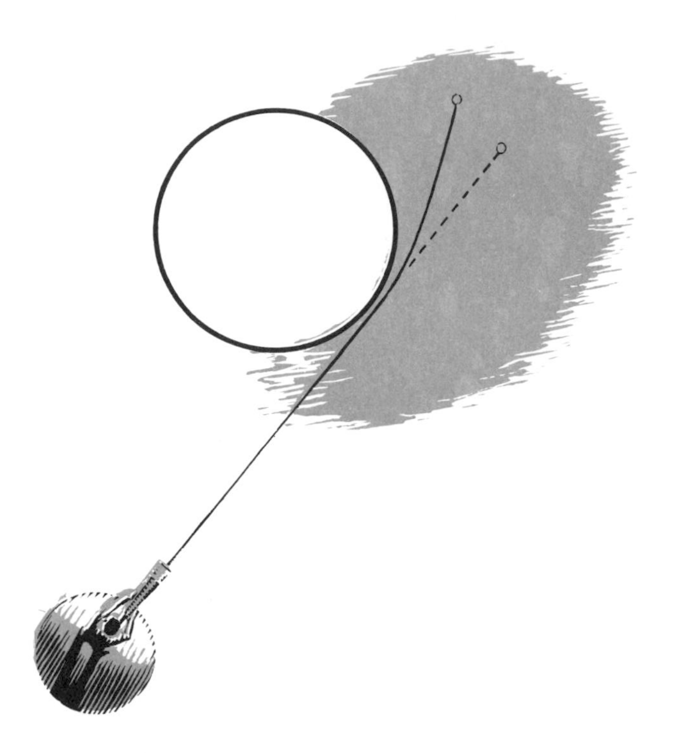

"And what," said one of them, "if we get a deflection twice as big as Einstein predicted?"

"Then," said the other, "Eddington will go mad."

Happily, the deflection proved to be close to Einstein's prediction. It was the publicity surrounding this dramatic confirmation of general relativity that first brought the theory to the attention of the general public. Today, astronomers are skeptical of this confirmation. The difficulties in making precise measurements of star positions during an eclipse are much greater than Eddington supposed, and there have been differences in the results obtained during eclipses since 1919.

It is not hard to understand the reasons for these discrepancies. Measuring instruments have to be carried to the site of the total eclipse. This is usually in some remote region, such as the middle of an ocean, a desert, an arctic waste, or a swamp where alligators are snapping at the astronomers' legs. During the eclipse the temperature of the air drops suddenly, causing unpredictable changes in refraction that alter observed positions of stars. Control pictures, showing the same star pattern when the sun isn't there, either have to be made on the spot, many months later when atmospheric conditions have changed, or compared with photographs taken at observatories in other regions. No wonder the results scatter as much as they do, and we have not even considered the influence of unconscious bias on the part of astronomers who have preconceived ideas of what kind of data they are expected to find.

Einstein's theory predicts an average deflection of 1.75 seconds of arc for each star. Two measurements in 1919 showed deflections of 1.98 and 1.61, both fairly close. But the deflection dropped to 1.18 in a 1922 test and rose to 2.24 in a 1929 test. At a 1962 meeting of the Royal Society of London, a group of scientists concluded that the difficulties are so great that eclipse observers should no longer attempt such measurements.

Dozens of sophisticated tests of general relativity in the 1970s supported Einstein's theory and tended to discredit all its rivals. Spacecrafts orbiting the sun have exchanged radio signals when they were positioned so that the waves would pass close to the sun and accurate measurements could be made of the time lag caused by the sun's gravitational field. Other tests have used radar signals from earth that passed close to the sun as they bounced off Mercury and Venus. Still other tests involve radio emissions coming from distant quasars (see Chapter 11) and passing close to the sun's rim. New tests are being made almost every month, and Einstein's theory of gravity is passing all of them.

A careful test, reported in 1976, was based on the round-trip travel times of laser beams aimed at reflectors left on the moon by astronauts. From these data, fantastically accurate measurements of the moon's motion can be made. Dicke's theory predicts a deviation of several feet from what is predicted by general relativity. No departure from Einstein's equations could be detected within the limits (about five inches) of measurement.

A scientific theory is not very powerful unless one can think of experiments that might strongly refute it. George Gamow, the eminent Russian-born physicist who died in 1968, described one such experiment involving antiparticles.* In 1957 Philip Morrison and Thomas Gold conjectured that antiparticles may have negative gravitational mass. If so, any gravitational force acting upon them would cause them to accelerate in a negative direction. An antiapple made of antimatter would have flown up in the sky instead of falling on Newton's nose. The conjecture is attractive because, if true, it would explain the absence of antimatter in our galaxy. Any antimatter produced in the past, in the vicinity of the galaxy, would long ago have been projected outward. Whether antiparticles have negative gravitational mass has not yet been unequivocally determined, but if it is found that they do, relativity theory will be in serious trouble.

To understand why there would be a difficulty, imagine a spaceship suspended in interstellar space, motionless with respect to the stars. Floating inside the spaceship is a solitary antiapple with negative gravitational mass.

* See Gamow's article, "Gravity," in *Scientific American* (March 1961).

The ship starts to move in the direction of the ceiling with an acceleration of one *g*. (A "*g*" is the acceleration with which bodies fall to the earth.)

What happens to the apple?

From the standpoint of an observer outside the ship, attached to the inertial frame of the cosmos, the apple should stay right where it is relative to the stars. No force is acting on it. The ship itself does not touch the apple; the ship might just as well be a thousand miles away. The floor of the compartment should, therefore, move up until it hits the apple. (We don't have to worry, in this thought experiment, about what happens when the floor hits the apple.)

The situation is altogether different if the ship is taken as a fixed frame of reference. Now the observer must suppose a gravitational field acting inside the ship. This would send the apple toward the ceiling with an acceleration (relative to the stars) of two *g*. A basic principle of relativity has been violated: The two frames of reference are not interchangeable.

In other words, negative gravitational mass is difficult to reconcile with general relativity, although Newton's approach to inertia accommodates it easily. Classical physics simply takes the first point of view. The ship has an absolute motion with respect to the ether. The apple remains at absolute rest. No gravitational field enters to complicate the picture.

It is important to realize that relativity is in trouble only when one type of mass is positive, the other negative. A body with both masses negative introduces no contradictions, although its behavior would be surprising.* A moving baseball made of such matter could be caught in a glove only by slapping it hard in the direction in which it is moving. However, the ball would fall to

* See Hermann Bondi, "Negative Mass in General Relativity," *Reviews of Modern Physics* (July 1957), and Banesh Hoffmann, "Negative Mass," *Science Journal* (April 1965).

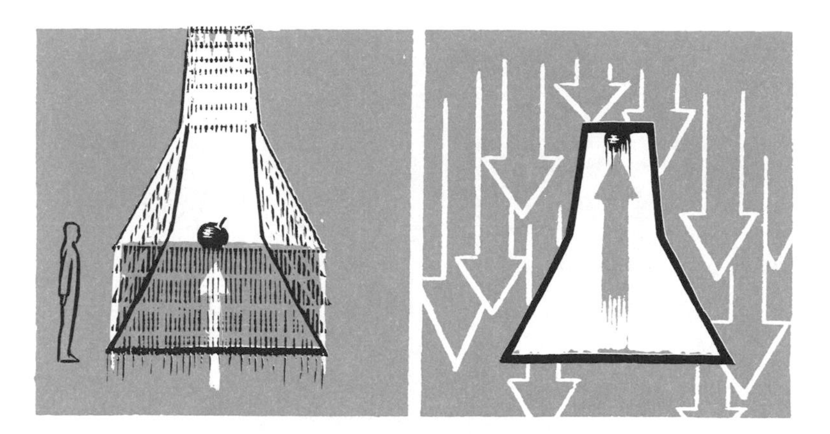

earth in the normal manner. In the case of a falling ball made of ordinary matter, the gravitational attraction between it and the earth is balanced by the drag of its inertial mass. In the case of a falling ball with both masses negative, the gravitational *repulsion* between it and the earth is balanced by the *pull* toward the earth of its inertial mass. Therefore it, too, falls normally in the earth's gravitational field.

Assume that "positive mass" means both masses are positive, and "negative mass" means both masses are negative. A positive mass attracts both kinds of masses, a negative mass repels both kinds. If a positive-mass star were in the vicinity of a negative-mass star, the positive star would attract, the negative star would repel. As a result, one star would chase the other through space with uniform acceleration! The negative star would gain in negative energy as its speed increased, otherwise the system would increase in energy in violation of the law of conservation of mass-energy.

The discovery of a particle with positive inertial mass and negative gravitational mass would be, as we have seen, altogether different. It would introduce a contradiction that would be fatal to general relativity and force a return to Newton's view of inertia as arising from absolute motion with respect to a fixed space. "The author earnestly hopes," Gamow concludes, "that this will not come to pass."

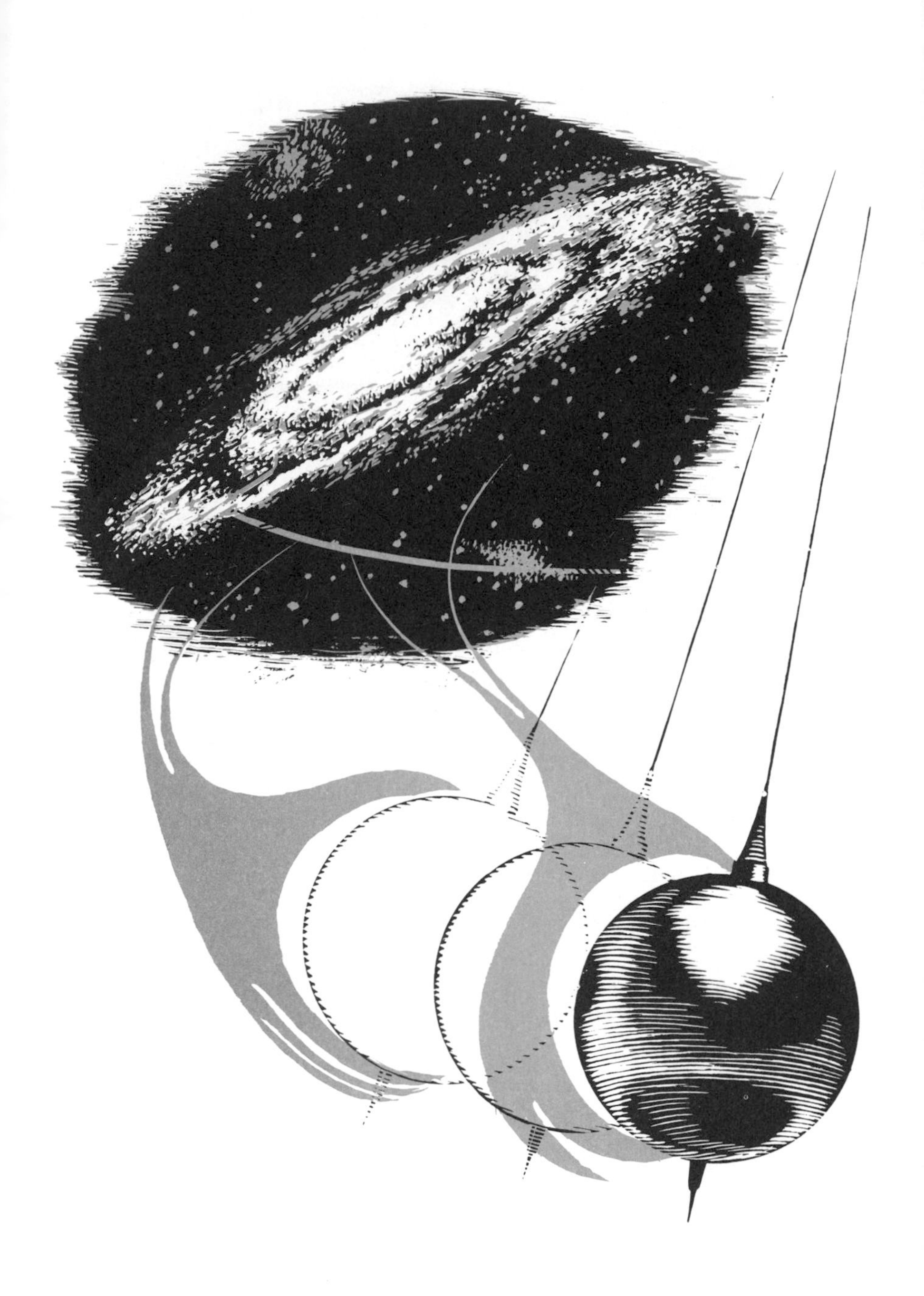

8

Mach's Principle

Einstein's principle of equivalence says that when an object is accelerated or rotated, a force field is created which can be looked upon either as inertial or gravitational, depending on the choice of a reference frame. A question of great importance now arises; a question that leads quickly into deep, yet-unsolved problems.

Are these force fields the result of motion with respect to a spacetime structure that exists independently of matter, or is the spacetime structure *created* by matter; that is to say, created by the galaxies and other material bodies of the universe?

Experts divide. All the old eighteenth-century and nineteenth-century arguments over whether "space" or the "ether" has an existence apart from matter are still with us; only now they are arguments about the spacetime structure (sometimes called the "metrical field") of the cosmos. Most of the early writers on relativity—Arthur Stanley Eddington, Bertrand Russell, Alfred North Whitehead, and others—believed that the structure is independent of the stars, though of course it is given local distortions by the stars. More plainly, if there were no other objects in the cosmos except the earth, it would still be possible, so these writers contended, for the earth to rotate relative to this spacetime structure. (It is irrelevant to this argument whether the structure has an overall positive, negative, or zero curvature.) A lone spaceship, the sole object in the universe, could still turn on its rocket motors and accelerate. Inside the ship, astronauts would still feel the inertial forces of acceleration. A lone earth, rotating in space, would still bulge around its middle. It would bulge because particles of its matter would be forced into paths that were not geodesics in the spacetime structure. The particles would go, so to speak, against the natural "grain" of spacetime. It would even be possible, on such a lone earth, to measure a type of inertial force called the Coriolis force* and determine the *direction* in which the earth was spinning.

Einstein granted the possible truth of this view, but he did not (at least as a young man) find it to his taste. He preferred instead a point of view that had first been advanced by the Irish philosopher Bishop Berkeley. If the earth were the only body in the universe, Berkeley argued, it would be meaningless to say that it could rotate. Somewhat similar views were

* If an intercontinental missile is traveling north or south, the rotation of the earth tends to deflect it to the right in the northern hemisphere, to the left in the southern hemisphere. This inertial effect is called the Coriolis force after G. G. Coriolis, an early nineteenth-century French engineer who was the first to analyze it completely. Cyclones and other circular movements of the atmosphere are traceable to Coriolis forces.

held in the seventeenth century by the German philosopher Gottfried von Leibniz and the Dutch physicist Christian Huygens, but it remained for Ernst Mach (the Austrian physicist mentioned in Chapter 2) to back up this view with a plausible scientific theory. Mach anticipated much of relativity theory, and Einstein has written about the extent to which Mach inspired his early thinking. (Sad to relate, Mach in his old age, after his insights had been incorporated by Einstein into a successful theory, refused to accept relativity.)

From Mach's point of view, a cosmos without stars would have no spacetime structure relative to which the earth could spin. For there to be gravitational (or inertial) fields capable of bulging a planet's equator and

spilling water over the sides of a rotating bucket, there must be stars to create a spacetime structure. Without such a structure, spacetime would possess no geodesics. It could not even be said that a light beam, speeding through completely empty space, would travel in a geodesic, because in the absence of a spacetime structure the beam would not know how to take one path rather than another. As expressed by one writer, A. d'Abro (in his classic work, *The Evolution of Scientific Thought*), it would not know which way to go. Even the existence of a spherical body such as the earth might be impossible. Particles of earth are packed together by gravity, and gravity moves particles along geodesics. With no spacetime structure and no geodesics, the earth (as d'Abro says) would not know what shape to take. Eddington once expressed this point humorously: in an entirely empty universe (if Mach is correct), Einstein's gravitational fields would fall to the ground!

D'Abro describes a thought experiment that helps clarify Mach's position. Imagine an astronaut floating in space. In his hand he holds a brick. There are no other objects in the universe. We know that the brick would have no weight (gravitational mass). Would it have inertial mass? If the astronaut tried to heave the brick into space, would it resist the movement of his hand? From Mach's point of view, it would not. With no stars in the cosmos to provide a metrical field for spacetime, there is nothing relative to which the brick can accelerate. Of course there is the astronaut, but his mass is so small that any effect relative to him would be negligible.

Einstein used the term "Mach's principle" for Mach's point of view. It was Einstein's early hope that this view could be incorporated into relativity theory. In fact, he once devised a model of the universe (to be discussed in Chapter 10) in which the spacetime structure of the universe has no existence

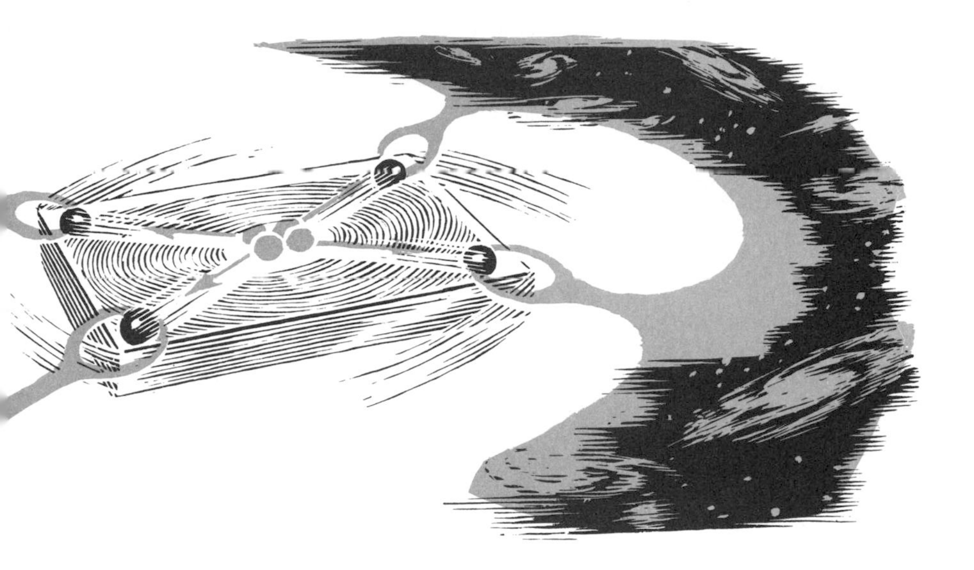

except insofar as it is created by the stars and other material bodies. "In a consistent theory of relativity," Einstein wrote in 1917 when he published his first mathematical description of this model, "there can be no inertia relative to 'space,' but only an inertia of masses relative to one another. If, therefore, I have a mass at a sufficient distance from all other masses in the universe, its inertia must fall to zero."

Later, serious flaws were discovered in Einstein's cosmic model and he was forced to abandon Mach's principle, but the principle continues to exert a strong fascination over today's cosmologists.* It is not difficult to see why. It carries the relativity of motion to its ultimate. The opposing point of view, the view that assumes a spacetime metric even in the absence of stars, is really very close to the old ether theory. Instead of a motionless, invisible jelly called the ether, there is a motionless, invisible spacetime structure. By assuming this to be fixed, accelerations and rotations take on a suspiciously absolute character. In fact, proponents of this point of view have not hesitated to speak of rotations and accelerations as "absolute." But if inertial effects are relative not to such a structure but only to a structure generated by the stars, then a very pure form of relativity is preserved.

Dennis Sciama, a British cosmologist, has developed an ingenious theory along Machian lines. He gives an entertaining account of it in his popularly written book, *The Unity of the Universe.* According to Sciama, inertial effects due to rotation or acceleration are the result of a relative motion with respect to the total matter in the universe. If this is true, then a measurement of inertia provides a method for estimating the amount of matter in the universe! Sciama's equations show that the influence of nearby stars on inertia is astonishingly small. All the stars in our galaxy, he believes, contribute only about one ten-millionth of the strength of inertia on the earth. Most of its strength is contributed by distant galaxies. Sciama estimates that 80 percent of inertial force is the result of motion relative to galaxies so distant that they have not yet been discovered by our telescopes!

In Mach's day it was not known that galaxies other than our own existed, nor was it known that our galaxy rotates. Astronomers today know that centrifugal force, arising from rotation, causes our galaxy to bulge enormously. From Mach's point of view this bulge could occur only if vast quantities of matter existed outside the galaxy. Had Mach known of the inertial effect of rotation on our galaxy, Sciama points out, he would have been able to deduce the existence of other galaxies fifty years before any of them were discovered.

The startling character of Sciama's point of view can be made even more evident by the following illustration. I once owned a small glass-topped puzzle, shaped like a square and containing four steel balls. Each ball rested on a groove that ran from the square's center to one of its corners. The problem was to get all four balls into the corners at the same time. The only way to solve it was by placing the puzzle flat on a table and spinning it. Centrifugal force did the trick. If Sciama is right, this puzzle could not be solved in this way if it were not for the existence of billions of galaxies at enormous distances from our own.

* For an excellent discussion of current attitudes toward Mach's principle, and various interpretations placed upon it, see R. H. Dicke's paper, "The Many Faces of Mach," in *Gravitation and Relativity*, edited by Hong-yee Chiu and William F. Hoffman (New York: The Benjamin Co., 1964).

Will the future of physics move in the direction of Mach, or will it retain an absolute structure of spacetime, independent of matter and waves? No one can say. If a successful field theory is developed, in which particles and their fields can be explained in terms of spacetime structure, then the stars will become merely one aspect of the geometry of spacetime. Instead of stars generating the spacetime structure, the structure will generate the stars.

These are deep questions. Physicists are nowhere near answering them.

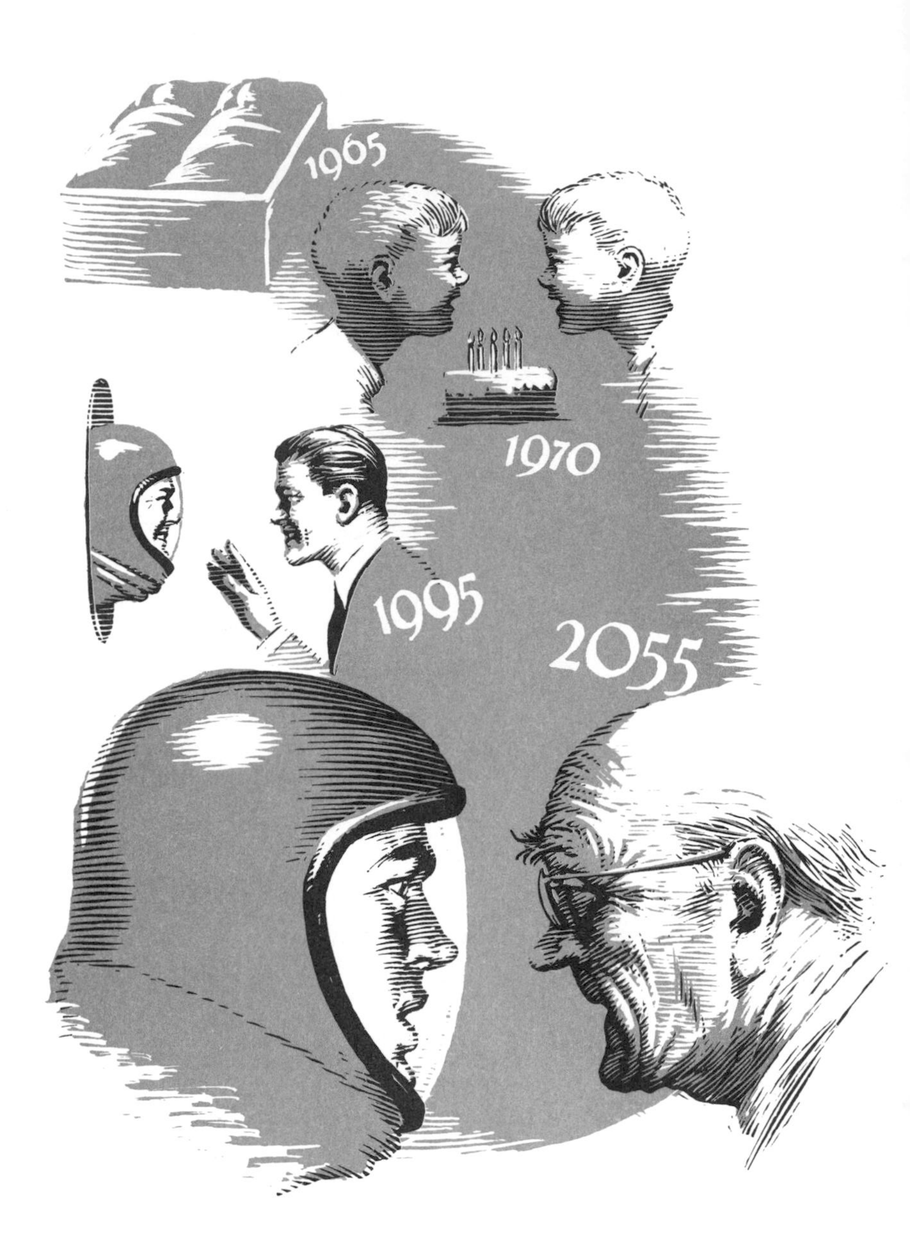
1965
1970
1995
2055

9 The Twin Paradox

How did the world's leading scientists and philosophers react when they caught their first glimpse of the strange new world of relativity? The reaction was mixed. Most physicists and astronomers, confused by the violations of common sense and the difficult mathematics of the general theory, maintained a discreet silence. But scientists and philosophers capable of understanding relativity were inclined to accept it with exhilaration. It has already been mentioned how quickly Eddington perceived the greatness of Einstein's achievement. Moritz Schlick, Bertrand Russell, Rudolf Carnap, Ernst Cassirer, Alfred North Whitehead, Hans Reichenbach, and many other eminent philosophers were early enthusiasts who wrote about the theory and tried to clarify its implications. Russell's book *The ABC of Relativity*, first published in 1925, is still one of the best popular accounts of relativity ever written.

Here and there scientists were unable to shake themselves loose from old Newtonian habits of thought. In many ways they resembled the scientists back in the days of Galileo who could not bring themselves to admit that Aristotle might have been mistaken. Michelson himself, a limited mathematician, never accepted relativity, even though his great experiment smoothed the way for the special theory. As late as 1935, when I was an undergraduate at the University of Chicago, I took a course in astronomy from Professor William D. Macmillan, a widely respected scientist. He was openly scornful of relativity.

"We of the present generation are too impatient to wait for anything," Macmillan wrote in 1927. "Within forty years of Michelson's failure to detect the expected motion of the earth with respect to the ether we have wiped out the slate, made a postulate that by no means whatever can the thing be done, and constructed a non-Newtonian mechanics to fit the postulate. The success which has been attained is a marvelous tribute to our intellectual activity and our ingenuity, but I am not so sure with respect to our judgment."*

All sorts of objections were raised against relativity. One of the earliest, most persistent objections centered around a paradox that had first been mentioned in 1905 by Einstein himself, in his paper on special relativity. (The word "paradox" is used in the sense of something opposed to common sense, not something logically contradictory.) This paradox is very much in the scientific news today because advances in space flight, coupled with progress in building fantastically accurate timing devices, may soon provide a way to test the paradox in a very direct manner.

The paradox is usually described as a thought experiment involving twins. They synchronize their watches. One twin gets into a spaceship and makes a long trip though space. When he returns, the twins compare watches. According to the special theory of relativity, the traveler's watch will show a slightly earlier time. In other words, time on the spaceship will have gone at a slower rate than time on the earth. So long as the space journey is confined to the solar system, and made at relatively low speeds, this time difference will be negligible. But over long distances, with velocities close to that of light, the "time dilation" (as it is sometimes called) can be large. It is not inconceivable that someday a means will be found by which a spaceship can be slowly accelerated until it reaches a speed only a trifle below that of light. This would make possible visits to other stars in the galaxy, perhaps even trips to other galaxies. So, the twin paradox is more than just a parlor puzzle; someday it may become a common experience of space travelers.

Suppose that the astronaut twin goes a distance of a thousand light-years and returns: a small distance compared with the diameter of our galaxy. Would not the astronaut surely die long before he completes the trip? Would

* From Macmillan's contribution to *A Debate on the Theory of Relativity*, by Robert D. Carmichael and others (La Salle, Ill.: Open Court Publishing Company, 1927).

not his trip require, as in so many science-fiction stories, an entire colony of men and women so that generations would live and die while the ship was making its long interstellar voyage?

The answer depends on how fast the ship goes. If it travels just under the limiting speed of light, time within the ship will proceed at a much slower rate. Judged by earth-time, the trip will take more than two thousand years. Judged by the astronaut on the ship, if he travels fast enough, the trip may take only a few decades!

For readers who like specific figures, here is a recent calculation by Edwin M. McMillan, a nuclear physicist at the University of California in Berkeley. An astronaut travels from the earth to the spiral nebula in Andromeda. Assume that the nebula is 1.5 million light-years from the earth (a conservative estimate; some astronomers believe it is closer to 2 million) and that the ship travels at such speed that the astronaut ages fifty-five years while making the trip there and back. When he returns, he finds that on the earth 3 million years have gone by!

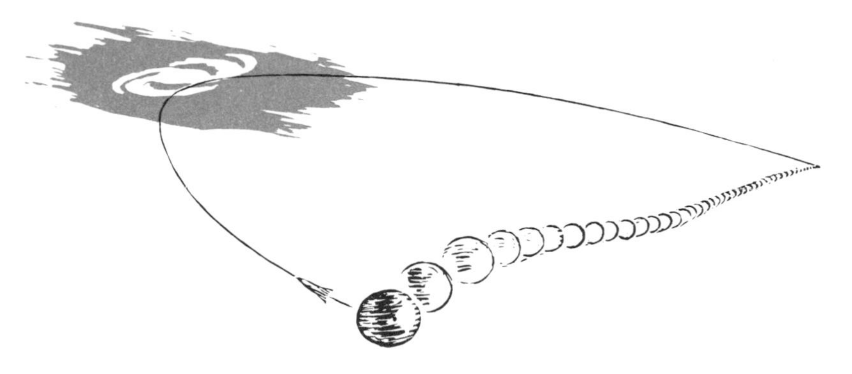

(Note of caution: The reader should regard all references in this book to interstellar or intergalactic space trips, made at speeds close to that of light, as primarily thought experiments intended to clarify aspects of relativity. For a good account of the enormous practical difficulties in obtaining such speeds, see Edward Purcell's contribution to *Interstellar Communication*, edited by A. G. W. Cameron (Benjamin, 1963). "All this stuff about traveling around the universe in space suits," Purcell concludes, "—except for *local* exploration, which I have not discussed—belongs back where it came from, on the cereal box." Perhaps.)

You can see at once that this raises all sorts of fascinating possibilities. A scientist of forty and his teen-age laboratory assistant fall in love. They feel that their age difference makes a marriage out of the question. So off he goes on a long space voyage, traveling close to the speed of light. He returns, age forty-one. Meanwhile, on the earth his girl friend has become a woman of thirty-three. Perhaps she could not wait fifteen years for her lover to return; she has married someone else. The scientist cannot bear this. Off he goes on another long trip. Moreover, he is curious to know if a certain theory he has published is going to be confirmed or discarded by later generations. He returns to earth, age forty-two. His former girl friend is long since dead. What is worse, his pet theory has been demolished. Humiliated, he takes an even longer trip, returning at the age of forty-five to see what the world is like a few thousand years hence. Perhaps, like the time traveler in H. G. Wells's story "The Time Machine," he will find that humanity has become obsolete. Now he is stranded. Wells's time machine could go both ways, but our lonely scientist has no means of getting back into the stream of human history where he belongs.

Unusual moral questions would arise if this sort of time travel became possible. Is there anything wrong, for instance, in a girl marrying her own great-great-great-great-great-great-grandson?

Please note: This kind of time travel avoids all the logical traps that plague science fiction, such as dropping into the past to kill your parents before you are born, or whisking into the future and shooting yourself between the eyes. Consider, for example, the plight of Miss Bright in that familiar limerick:

> There was a young lady named Bright,
> Who traveled much faster than light.
> She started one day
> In the relative way,
> And returned on the previous night.*

If she returned on the previous night, then she must have encountered a duplicate of herself. Otherwise it would not have been truly the night before. But there could not have been two Miss Brights the night before because the time-traveling Miss Bright left with no memory of having met her duplicate yesterday. So you see, there is a clear-cut contradiction. Time travel of *that* sort is not logically possible unless the existence of parallel worlds running along branching time tracks is assumed. Even with this gimmick, matters become quite complicated.

Note also that Einstein's form of time travel does not confer upon the traveler any genuine immortality, or even longevity. As far as *he* is concerned, he always ages at the normal rate. It is only the earth's "proper time" that for the traveler seems to gallop along at breakneck speed.

Henri Bergson, the famous French philosopher, was the most eminent

* This limerick about Miss Bright was written by A. H. Reginald Buller, a professor of botany at the University of Manitoba, and first published in *Punch.* The contradiction that arises from Miss Bright's journey back in time also applies to "tachyons" (conjectured particles that go faster than light) if tachyons can be used for transmitting signals. See my column on time travel in *Scientific American* (May 1974).

thinker to cross swords with Einstein over the twin paradox. He wrote about it at some length, poking fun at what he thought were its logical absurdities. Unfortunately, what he wrote only proves that it is possible to be a great philosopher without knowing much about mathematics.* In the 1970s the same objections were raised again. Herbert Dingle, an English physicist, refused to believe the paradox. For years he wrote witty articles about it, accusing other relativity experts of being either obtuse or evasive. The superficial analysis to be given here will not clear up this controversy, which quickly plunges into complicated equations, but it will explain in a general way why there is almost universal agreement among experts that the twin paradox will really carry through in just the manner Einstein described.

Dingle's objection, the strongest that can be made against the paradox, is stated this way. According to the general theory of relativity, there is no absolute motion of any sort, no "preferred" frame of reference. It is always possible to choose a moving object as a fixed frame of reference without doing violence to any natural law. When the earth is chosen as a frame, the astro-

* Bergson's attack is in his book *Durée et Simultanéité* (3rd ed.; Paris, 1926). In the United States the same naïve arguments were repeated by philosophers William Pepperell Montague and Arthur Oncken Lovejoy. See Montague's "The Einstein Theory and a Possible Alternative," *Philosophical Review*, Vol. 33 (March 1924), pages 143–170. (Montague's alternative is the assumption—physicists now call it the Ritz theory—that light *is* influenced by the motion of its source; his attacks on Einstein reveal an amazing lack of comprehension of relativity theory.) For Lovejoy's attack on the twin paradox, see "The Paradox of the Time-Retarding Journey, Part I," *Philosophical Review*, Vol. 40 (January 1931), pages 48–68; Part II appeared in the March issue, same volume, pages 152–167. Lovejoy concludes that Bergson is right: there are many "fictitious times" but "only one real time." Evander Bradley McGilvary rebuts Lovejoy in the July issue, pages 358–379, but Lovejoy, unconvinced, replies in November, pages 549–567.

naut makes the long journey, returns, finds himself younger than his stay-at-home brother. All well and good. But what happens when the spaceship is taken as the frame of reference? Now it must be assumed that the earth makes a long journey away from the ship and back again. In this case it is the twin on the ship who is the stay-at-home. When the earth gets back to the spaceship, will not the earth rider be the younger? If so, the situation is more than a paradoxical affront to common sense; it is a flat logical contradiction. Clearly, each twin cannot be younger than the other.

Dingle likes to state it this way: Either the assumption must be made that after the trip the twins will be exactly the same age or relativity must be discarded.

Without going into any of the actual computations, it is not hard to understand why the alternatives are not so drastic as Dingle would have us believe. It is true that all motion is relative, but in this case there is one all-important difference between the relative motion of the astronaut and the relative motion of the stay-at-home. *The stay-at-home does not move relative to the universe.*

How does this affect the paradox?

Assume that the astronaut is off to visit Planet X, somewhere in the galaxy. He travels at a constant speed. The stay-at-home's watch is attached to the inertial frame of the earth, on which there is agreement among clocks because they are all relatively motionless with respect to each other. The astronaut's watch is attached to a different inertial frame, the frame of the ship. If the ship just kept on going forever, there would be no paradox because there would be no way to compare the two watches. But the ship has to stop and turn around at Planet X. When it does so, there is a change from an inertial frame moving away from the earth to a new inertial frame moving toward the earth. This shift is accompanied by enormous inertial forces as the ship accelerates during the turnaround. In fact, if the acceleration during the turnaround were too great, the astronaut (and not his twin on the earth) would be killed. These inertial forces arise, of course, because the astronaut is accelerating with respect to the universe. They do not arise on the earth, because the earth is not undergoing similar acceleration.

From one point of view it can be said that the inertial forces produced by this acceleration "cause" a slowing down of the astronaut's watch; from another point of view the acceleration merely indicates a shift of inertial frames. Because of this shift, the world line of the spaceship—its path when plotted on Minkowski's four-dimensional graph of spacetime—becomes a path on which the total "proper time" of the round trip is less than the total proper time along the world line of the stay-at-home twin.* Although acceleration is involved in the shifting of inertial frames, the actual computation involves nothing more than the equations of the special theory.

* To see exactly how this works out mathematically, read the excellent article on "The Clock Paradox in Relativity Theory," by Alfred Schild, in *American Mathematical Monthly* (January 1959).

Dingle's objection still remains, however, because exactly the same calculations can be made by supposing that the spaceship instead of the earth is the fixed frame of reference. Now it is the earth that moves away, shifts inertial frames, comes back again. Why wouldn't the same calculations, with the same equations, show that earth-time slowed down the same way? They would indeed if it were not for one gigantic fact: when the earth moves away, *the entire universe moves with it.* When the earth executes its turnaround, the universe does also. This accelerating universe generates a powerful gravitational field. As explained earlier, gravity has a slowing effect on clocks. A clock on the sun, for instance, would tick more slowly than the same clock on earth, more slowly on the earth than on the moon. Now, it turns out, when all the proper calculations are made, that the gravitational field generated by the accelerating cosmos slows down the spaceship clocks until they differ from earth clocks by precisely the same amount as before. This gravity field has, of course, no effect on earth clocks. The earth does not move relative to the cosmos; therefore, there is no gravitational field with respect to the earth.

It is instructive to imagine a situation in which the same time difference results, even though no accelerations are involved. Spaceship A passes the earth with uniform speed, on its way to Planet X. As the ship passes the earth it sets its clock at zero time. Ship A continues with uniform velocity to Planet X, where it passes spaceship B, moving with uniform speed in the opposite direction. As the ships pass, A radios to B the amount of time (measured by its own clock) that has elapsed since it passed the earth. Ship B notes this information and continues with uniform speed to the earth. As it passes the earth it radios to the earth the length of time A took to make the trip from the earth to Planet X, together with the length of time it took B (measured by its own clock) to make the trip from Planet X to earth. The total of these two periods of time will be less than the time (measured by earth clocks) that has elapsed between the moment that ship A passed the earth and the moment that ship B passed the earth.

This difference in time can be calculated by the equations of the special theory. No accelerations of any sort are involved. Of course, now there is no twin paradox because there is no astronaut who goes out and comes back. It can be supposed that the traveling twin rides out on ship A, then transfers to ship B and rides back, but there is no way he can do this without transferring from one inertial frame to another. To make the transfer he must undergo incredibly strong inertial forces. These forces indicate his shift of inertial frames. If we wish, we can say that the inertial forces slow down his clock. However, if the whole episode is viewed from the standpoint of the traveling twin, taking him as the fixed frame of reference, then a shifting cosmos that sets up gravitational fields enters the picture. (A major source of confusion in discussing the twin paradox is that the situation can be described in so many different verbal ways.) Regardless of the point of view adopted, the equations of relativity give the same time difference. This difference can be accounted for by the special theory alone. It is only to counter the objection raised by Dingle that the general theory must be brought into the picture.

It cannot be stated too often that it is not correct to ask which situation is "right": Does the traveling twin move out and back or do the stay-at-home and the cosmos move out and back? There is only *one* situation: a relative motion of the twins. There are, however, two different ways of talking about it. In one language, a change of inertial frames on the part of the astronaut, with its resulting inertial forces, accounts for the difference in aging. In the other language, gravitational forces overbalance the effect of a change of inertial frames on the part of the earth. *From either point of view, the stay-at-home and the cosmos do not move relative to one another.* Thus the situation is entirely different for each man, even though the relativity of motion is strictly preserved. The paradoxical difference in aging is accounted for, regardless of which twin is taken to be at rest. There is no need to discard the theory of relativity.

An interesting question can now be asked: What if the cosmos contained nothing except two spaceships, A and B? Ship A turns on its rocket engines, makes a long trip, comes back. Would the previously synchronized clocks on the two ships be the same?

The answer depends on whether you adopt Eddington's view of inertia or the Machian view of Dennis Sciama. In Eddington's view the answer is yes. Ship A accelerates with respect to the metric spacetime structure of the cosmos; ship B does not. The situation remains unsymmetrical and the usual difference in aging results. From Sciama's point of view the answer is no. Acceleration is meaningless except with respect to other material bodies. In this case, the only material bodies are the two spaceships. The situation is perfectly symmetrical. In fact, there are no inertial frames to speak of because there is no inertia (except an extremely feeble, negligible inertia resulting from the presence of the two ships). In a cosmos without inertia it is hard to predict what would happen if a ship turned on its rocket motors! As Sciama says, with British understatement, "Life would be quite different in such a universe."

Because the slowing of the traveling twin's time can be viewed as a gravitational effect, any experiment that shows a slowing of time by gravity provides a kind of indirect confirmation of the twin paradox. In recent years there have been several such confirmations by means of a wonderful new laboratory tool called the Mössbauer effect.* A young German physicist named Rudolf L. Mössbauer discovered, in 1958, how to make a "nuclear clock" that keeps unbelievably accurate time. Imagine one clock ticking five times every second and another clock ticking at so nearly the same rate that after a million million ticks it has lost only one hundredth of a tick. The Mössbauer effect is capable of detecting at once that the second clock is slower than the first! Experiments using the Mössbauer effect have shown that time near the bottom of a building (where gravity is stronger) is a bit slower than time near

* See Sergio DeBenedetti, "The Mössbauer Effect," *Scientific American* (March 1960).

the top of the same building. "A typist working on the first floor of the Empire State Building," Gamow observed, "will age slower than her twin sister working on the top floor." The difference in aging is, of course, infinitesimal; nevertheless, it is real and can be measured.

Physicists have also discovered, using the Mössbauer effect, that a nuclear clock slows down a bit when placed on the edge of a rapidly rotating disk as small as six inches in diameter. The revolving clock can be viewed as the traveling twin who undergoes constant changes of inertial frames (or alternatively, as the twin affected by a gravitational field if the disk is assumed at rest and the cosmos rotating), so this provides an excellent test of the twin paradox. The twin effect is also evident in the slower aging of muons, making circular trips in magnetic fields, as compared with muons that "stay at home."

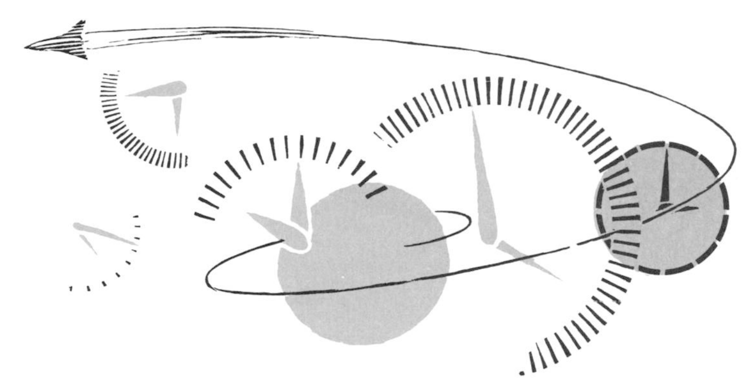

A more direct test was made in 1971 by Joseph Hafale and Richard Keating. They carried four atomic clocks around the earth on commercial jet liners, first circling the earth eastward, then making a western round trip. The eastward plane moved faster (relative to the universe) than the westward plane. Compared to a reference clock in Washington, the traveling clocks performed as expected. They lost time on the eastward trip, gained time on the westward trip. *Scientific American* (September 1972) called it the cheapest test ever made of relativity. It cost about $8,000, of which $7,600 was for air fare.

The time is rapidly approaching when an astronaut can make the final, definitive test by carrying a nuclear clock with him on a long space voyage. No physicist except Professor Dingle* doubts that the astronaut's clock, when he returns, will be slightly out of phase with a nuclear clock that stayed at home.

* Well, not quite. Dingle has a few supporters. An amusing history of the controversy, giving all sides and 305 references, is L. Marder's *Time and the Space-Traveller* (University of Pennsylvania Press, 1974). Dingle (who died in 1978) became persuaded that *all* of relativity, both special and general, is wrong. See his *Science at the Crossroads*, published by International Pubns. Service, 1974.

10 Models of the Universe

In this chapter the fairly solid, agreed-upon aspects of relativity are left behind and the reader is plunged into a misty region of strong controversy: a region where views are no more than tentative suggestions to be accepted or rejected on the basis of evidence that science does not yet possess. What is the universe like as a whole? We know that the earth is the third planet from the sun in a system of nine planets, and that the sun is one of about a hundred billion stars that make up our galaxy. We know that as far as the most powerful telescopes can probe, space is strewn with other galaxies, galaxies that also must be counted by the billions. Does this go on and on forever? Is there an infinity of galaxies? Or does the cosmos have a finite size?

Astronomers try to answer these questions as best they can by constructing what are called models of the universe: imaginary pictures of what the cosmos is like when viewed in its totality. In the early nineteenth century many astronomers assumed that the universe went on and on forever, containing an infinity of suns. Space was Euclidian. Straight lines extended to infinity in all directions. If a spaceship began a journey in any direction and continued in a straight line, it would go on endlessly without ever reaching a boundary. This, of course, is a view that goes back to the ancient Greeks. They liked to say that if a warrior kept throwing his spear farther and farther out into space, he could never reach an end; if such an end were imagined, the warrior could stand there and toss his spear still farther!

There is one important objection to this view. Heinrich Olbers, a German astronomer, pointed out in 1826 that if the number of suns is infinite, and the suns are randomly distributed in space, then a straight line from the earth, in any direction, would eventually intersect a star. This would mean that the entire night sky should be one solid, blinding expanse of starlight. Obviously, it isn't. Some explanation of the dark night sky has to be devised to explain what is now called Olbers' paradox. Most astronomers of the late nineteenth

and early twentieth centuries explained it by saying that the number of suns is finite. Our galaxy, they argued, contains all the suns there are. Outside the galaxy? Nothing! (It was not until the mid-twenties of this century that the evidence became overwhelming that there were millions of galaxies at enormous distances from our own.) Other astronomers suggested that the light of distant stars may be blotted out by masses of interstellar fog.

The cleverest explanation of all was advanced by the Swedish mathematician C. V. L. Charlier. Galaxies, he said, are grouped together in clusters. These clusters (he speculated) are grouped into superclusters, the superclusters into super-superclusters, and so on to infinity. At each step to a higher grouping, distances between the groupings grow larger in proportion to the sizes of the groups. If this were true, then the farther a straight line extended from our galaxy, the less the probability that it would encounter another galaxy. On the other hand, the hierarchy of clusters is endless, so it still can be said that the universe contains an infinity of stars. There is nothing wrong with Charlier's explanation of Olbers' paradox except that there is a simpler explanation. It will be given in a moment.

The first cosmic model based on relativity theory was proposed by Einstein himself in a paper published in 1917. It was an elegant, beautiful model, although Einstein later had to abandon it. As we learned earlier, gravitational fields are the warps or curves produced in the structure of spacetime by the presence of large masses of matter. Within every galaxy, therefore, there is a great deal of this twisting and bending of spacetime. What about the vast reaches of empty space between the galaxies? One point of view is that the farther space extends, away from the galaxies, the flatter (more Euclidian) space becomes. If the universe were empty of all matter, it would be completely flat, or perhaps it would be meaningless to say that it had any structure at all. In either case, the universe of spacetime stretches to infinity in all directions.

Einstein made an attractive counter-suggestion. Suppose, he said, the amount of matter in the universe is great enough to produce an overall positive curvature. Space would then curve back on itself in all directions. This cannot be fully understood without going into four-dimensional, non-Euclidian geometry, but the meaning can be grasped easily enough with the help of a two-dimensional model. Imagine a Flatland on which two-dimensional creatures live. They think of it as a Euclidian plane that extends to infinity in all directions. It is true that the suns of Flatland cause various bumps in the plane, but these are localized bumps that do not affect the overall flatness. There is, however, another possibility that might occur to Flatland astronomers. Perhaps each local bump produces a slight warping of the entire plane, so that the total effect of all the suns is to curve the plane until it becomes the surface of a bumpy sphere. Such a plane would still be boundless in the sense that you could move in any direction forever and never come to a boundary. A Flatland warrior would still be unable to find a spot

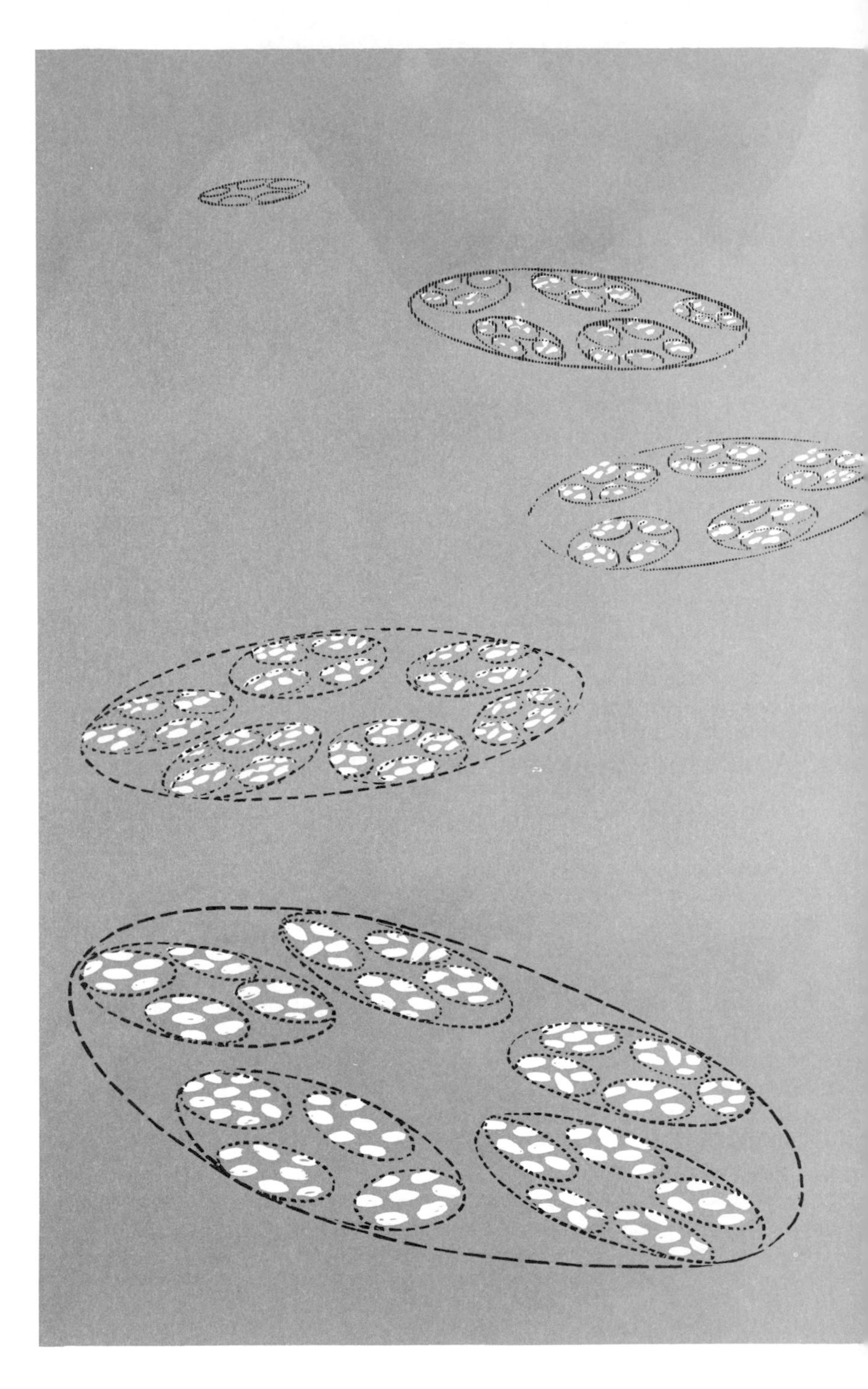

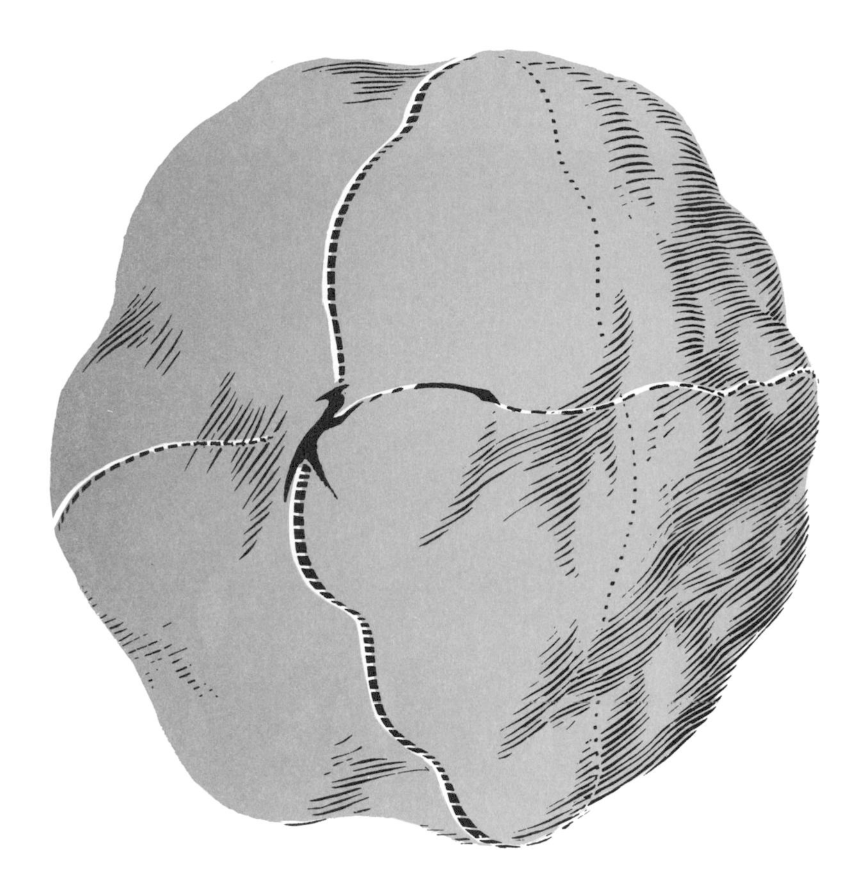

beyond which he could not toss his flat spear. Nevertheless, the surface would be finite. A trip continued long enough in a "straight line" would eventually bring the traveler back to where he started.

Mathematicians say that such a surface is "closed." It is finite but unbounded. Like infinite Euclidian space, its center is everywhere, its circumference nowhere. This "closure," a topological property of the surface, is one that Flatlanders could easily test. One test has already been mentioned: going around the sphere in all directions. Another test would be to paint the surface. If a Flatlander started at one spot and painted larger and larger circles, he would eventually paint himself *into* a spot on the opposite side of the sphere. If the sphere were large, however, and the Flatlanders confined to a small portion of its surface, they would be unable to make such topological tests.

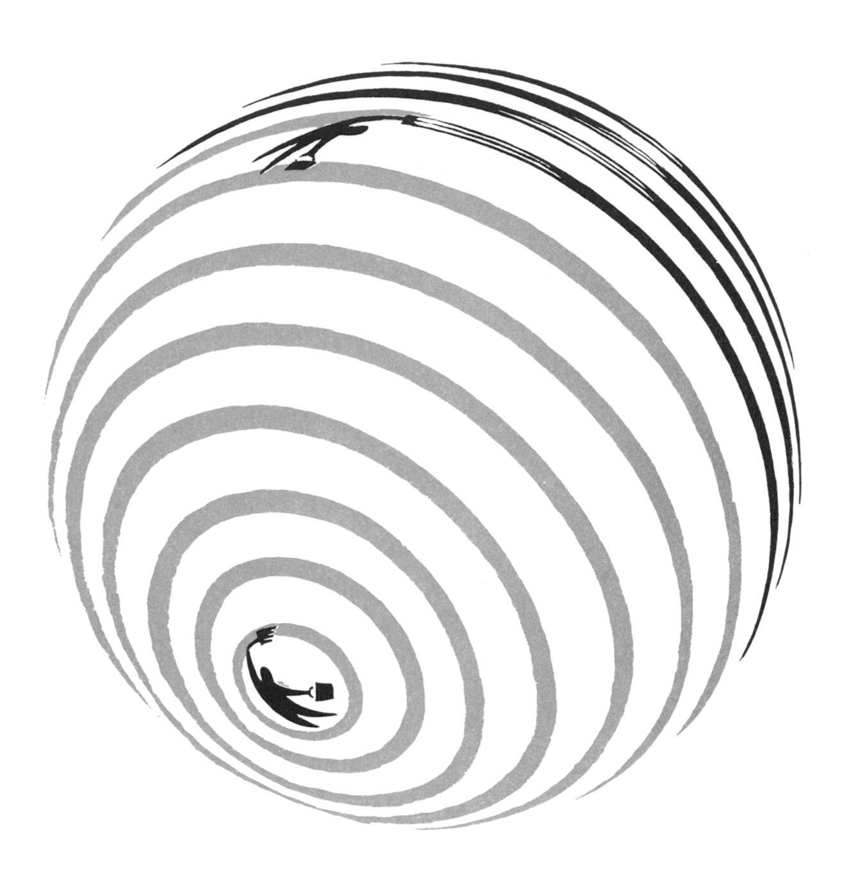

Einstein suggested that our space is the three-dimensional "surface" of a vast hypersphere (four-dimensional sphere). Time, in his model, remains uncurved: a straight coordinate extending back to an infinite past, forward to an infinite future. If the model is visualized as a four-dimensional spacetime structure, it is more like a hypercylinder than a hypersphere. For this reason, the model is usually called the "cylindrical universe." At any instant of time we see space as a kind of three-dimensional cross section of the hypercylinder. Each cross section is the surface of a hypersphere.

Our galaxy occupies only a minute portion of this surface, so it is not yet possible to perform a topological experiment that will prove its closure. A telescope powerful enough might be focused upon a certain galaxy in one direction, and then upon the *back* of the same galaxy by being pointed in the opposite direction. If there were spaceships that could approach the speed

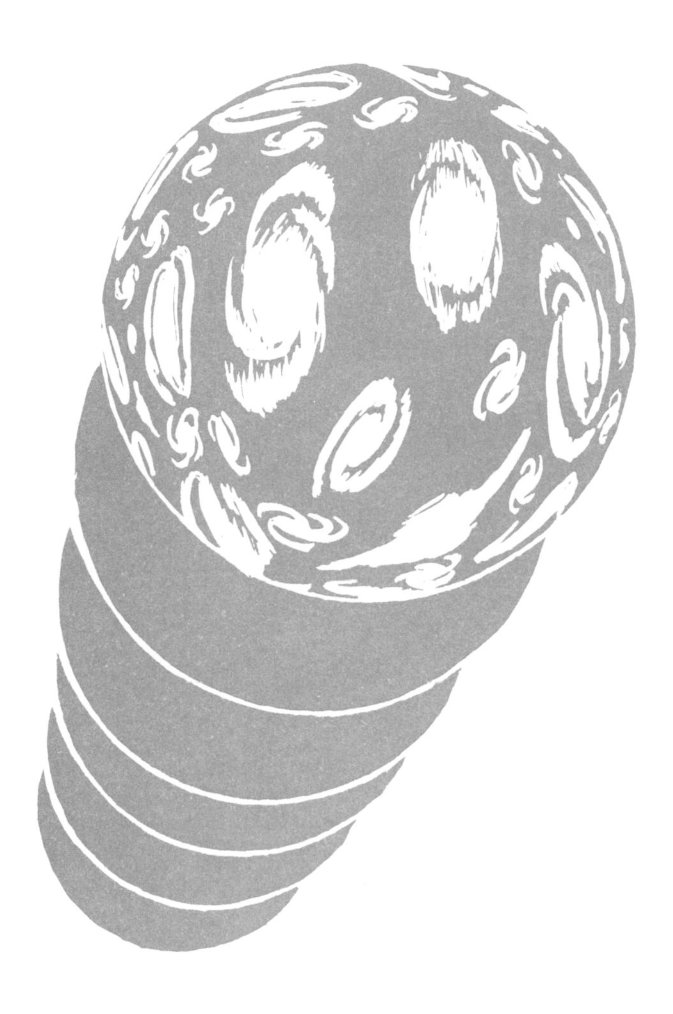

of light, they might be able to circle the cosmos by moving in any direction in the straightest possible line. The cosmos cannot literally be "painted," but essentially the same thing could be done by mapping it, making the spherical maps larger and larger. If the mapper continued long enough, he might find himself passing a point beyond which he would be *inside* the sphere he was mapping. This sphere would grow smaller and smaller as he continued mapping, like the circle that diminishes when a Flatlander paints himself into a spot.

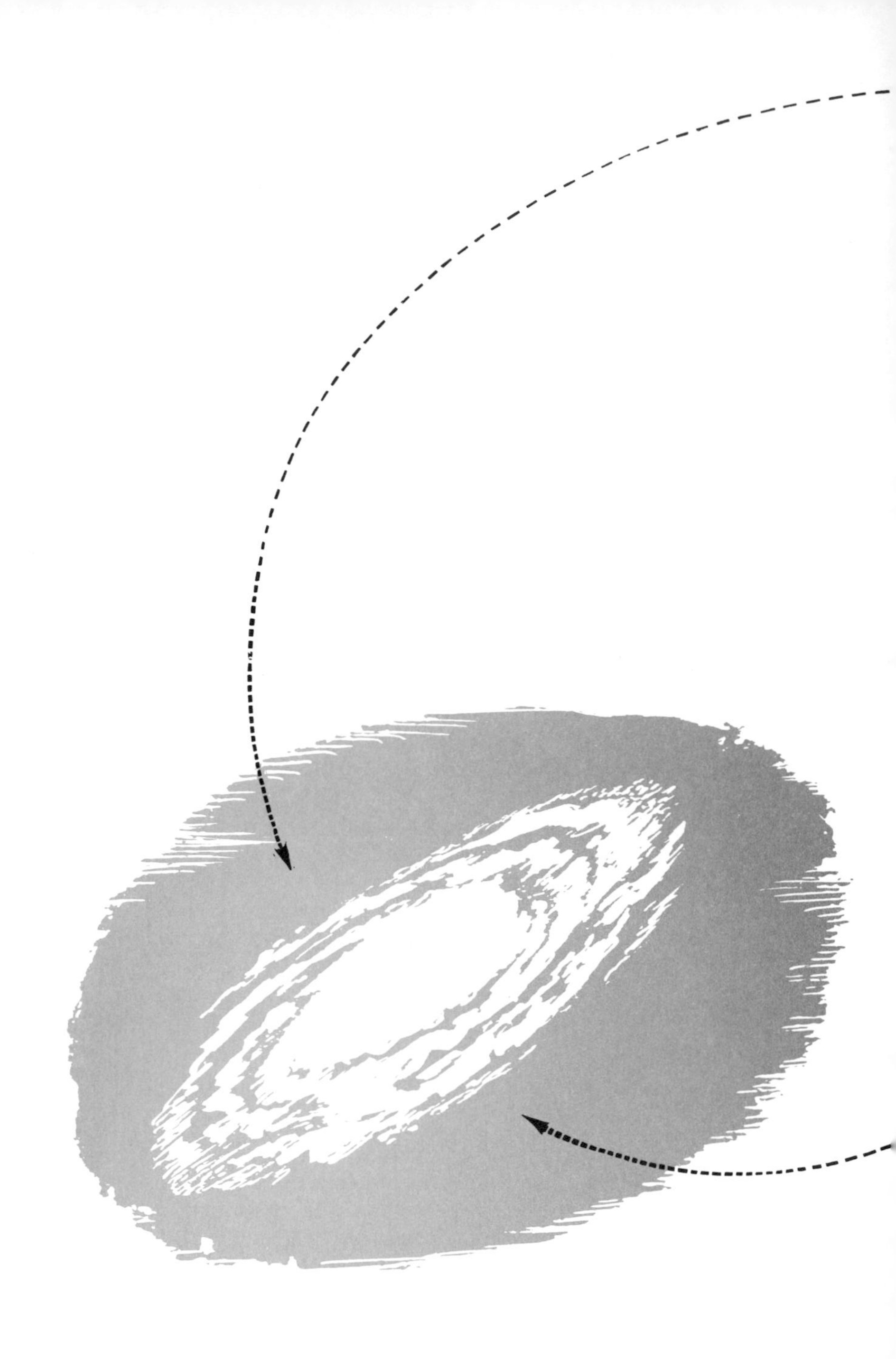

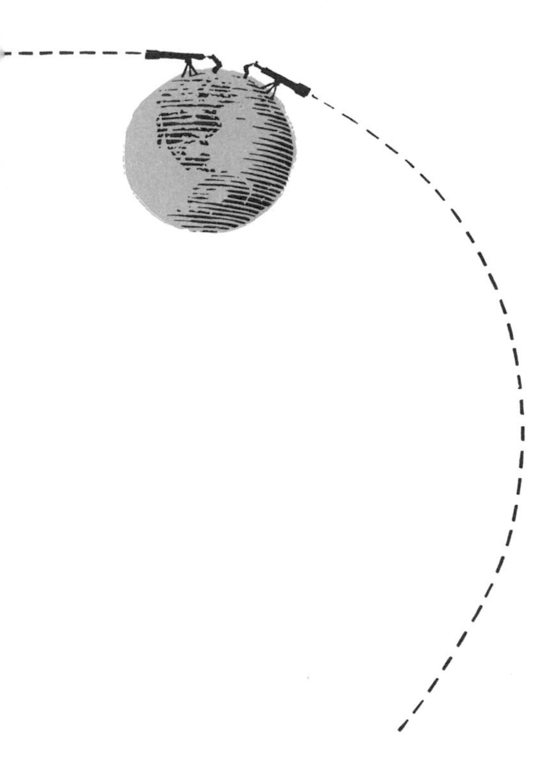

In some ways Einstein's non-Euclidian model is simpler than the classical model in which space is flat. It is simpler in the same sense that a circle may be said to be simpler than a straight line. A straight line stretches off to infinity at both ends, and infinity in mathematics is quite a complicated topic! A circle is comfortably finite. It has no ends; no one need worry about what happens to the line at infinity. Similarly, in Einstein's tidy universe no one need worry about all the loose ends at infinity, about what cosmologists like to call the "boundary conditions." There are no boundary problems in Einstein's cozy universe, because it has no boundaries.

Other cosmic models, all consistent with general relativity, were proposed and debated during the twenties. Some of them have properties even stranger than those of Einstein's cylindrical universe. The Dutch astronomer Willem de Sitter worked out another closed, finite model, but in his model, time curves as well as space. The farther one looks through de Sitter's space, the slower clocks seem to be running. If one looks far enough, he arrives at a region where time stops altogether, "like the Mad Hatter's tea party," writes Eddington, "where it is always six o'clock."

"Not that there is any boundary," explains Bertrand Russell in *The ABC of Relativity*. "The people who live in what our observer takes to be lotus land live just as bustling a life as he does, but get the impression that he is eternally standing still. As a matter of fact, you would never become aware of the lotus land, because it would take an infinite time for light to travel from it to you. You could become aware of places just short of it, but it would remain itself always just beyond your ken." Of course, if you traveled to this region in a spaceship, keeping the region under constant observation through a telescope, you would see its time slowly speeding up as you got nearer to it. When you arrived, everything would be moving at a normal rate. The lotus land would now be at the edge of a new horizon.

Have you ever noticed that when an airplane zooms low overhead the sound of its motors suddenly lowers a bit in pitch as the plane passes overhead? This is called the Doppler effect, after Christian Johann Doppler, an Austrian physicist who discovered the effect in the mid-nineteenth century. It is easily explained. As the plane approaches, its speed causes the pulses of sound from its engines to strike your eardrums at a faster rate than they would if the plane were not moving. This raises the pitch of the sound. As the plane moves away, the pulses strike your ears at a slower rate. The pitch lowers.

Exactly the same sort of thing happens when a light source moves rapidly toward or away from you. This has nothing to do with the velocity of light (which is always constant), but with the wavelengths of light. If you and a light source are in relative motion toward one another, the Doppler effect shortens the wavelength of light toward the violet end of the spectrum. If you and the light source are moving apart, the Doppler effect causes a similar shift toward the red end of the spectrum.

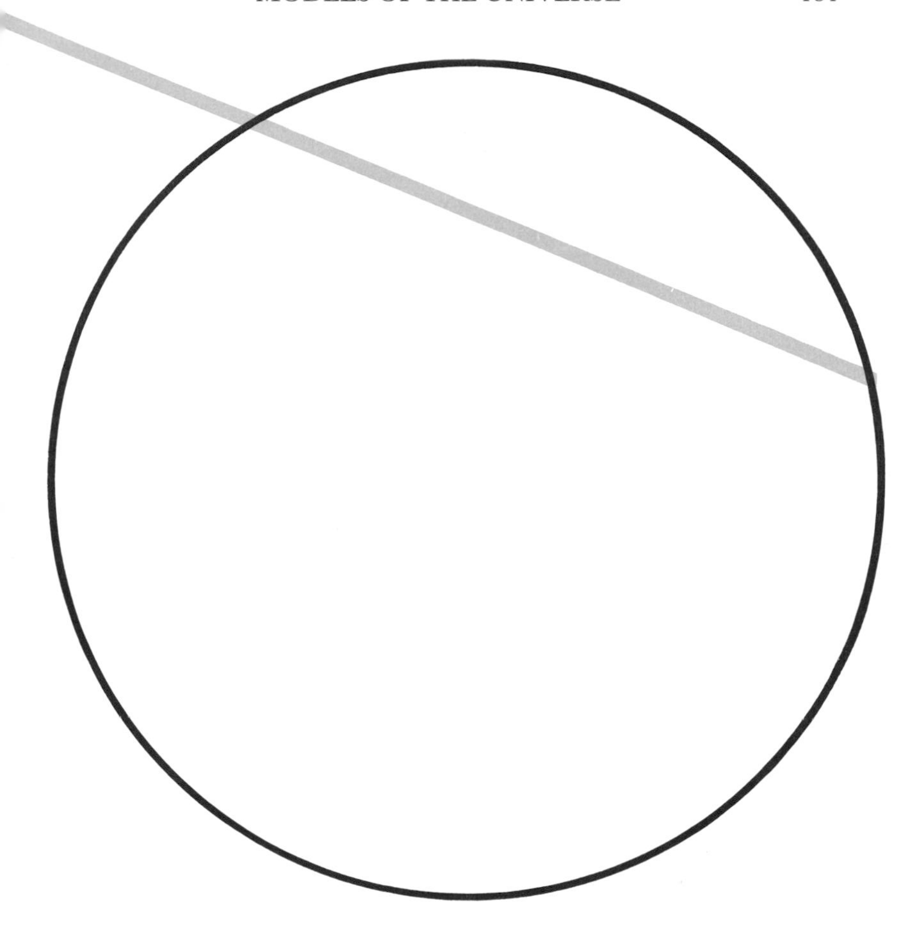

George Gamow, in one of his lectures, told a story (no doubt apocryphal) involving the Doppler effect that is much too good to be overlooked. It seems that Robert W. Wood, a famous American physicist at Johns Hopkins University, had been caught driving through a red light in Baltimore. In his appearance before the judge, Wood gave a brilliant account of the Doppler effect, explaining how his motion toward the red light had shifted the color toward the violet end of the spectrum, causing him to see it as green. The judge was set to waive the fine, but one of Wood's students (whom Wood had recently flunked) happened to be present. He pointed out the speed that would be required in order to shift the traffic light from red to green. The judge dropped the original charge, and fined Wood for speeding.

Doppler thought that the effect he discovered explained the apparent color of distant stars: Reddish stars would be moving away from the earth, bluish stars moving toward the earth. This turned out not to be the case (the colors have other causes), but during the 1920s it was discovered that light from distant galaxies shows a distinct shift toward the red that cannot be accounted for adequately except by assuming that the galaxies are moving away from the earth. Moreover, the shift increases, on the average, in the same proportion as the distance of the galaxy from the earth. If galaxy A is twice as far away as galaxy B, the redshift of A tends to be twice the redshift of B.

Various attempts have been made to account for this redshift in some other way than by assuming it to be a Doppler effect. One of them, the "tired light" theory, says simply that the longer light travels, the slower it vibrates. (This is a perfect example of an *ad hoc* hypothesis, because there is no other evidence to support it.) Another explanation is that the passage of light through cosmic dust causes the shift. De Sitter's model explains the shift neatly in terms of a curving time. But the simplest explanation, the one that fits best with other known facts, is that the redshift does indicate an actual motion of galaxies. A new series of "expanding universe" models were soon developed on this assumption.

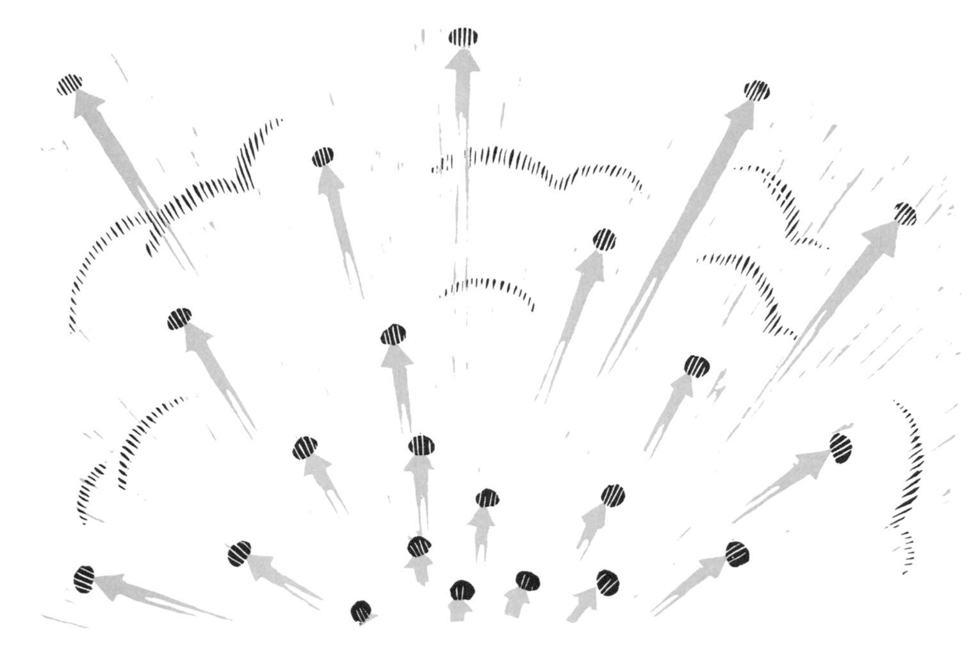

It is important to understand that this expansion does not mean that the solar system, or that galaxies, are expanding, or even (it now appears) that spaces between the galaxies in a galactic cluster are expanding. The expansion seems to involve only the spaces between the clusters. Imagine a huge lump of dough in which hundreds of raisins are embedded in a random way.

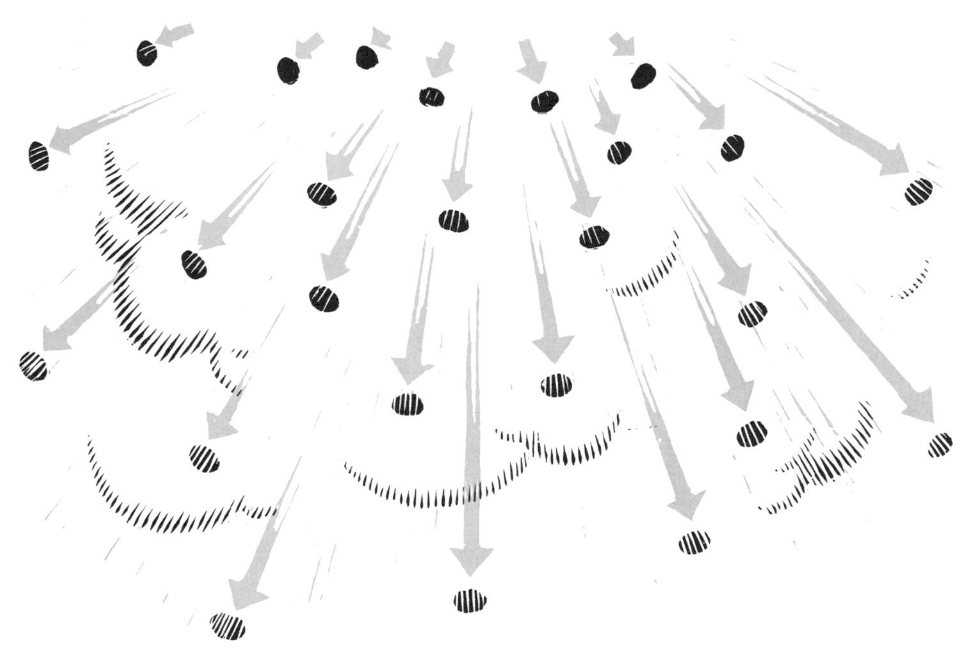

Each raisin represents a cluster of galaxies. If the dough is baked so that it expands uniformly in all directions, the raisins themselves remain the same size. It is the space between the raisins that grows larger. No one raisin can be called the center of this expansion. From the viewpoint of any individual raisin, all the other raisins seem to be moving away from it. The more distant the raisin, the faster it seems to recede.

Einstein's model of the universe is static. This, of course, is because he developed it before astronomers decided the universe was expanding. In order to prevent gravitational forces from pulling his cosmos together and collapsing it, he had to suppose that there is another force (he called it the "cosmological constant") which acts as a repelling influence and keeps the stars apart. Later calculations showed that Einstein's model is unstable, like a thin dime balanced on edge. The slightest shove would make it fall heads or tails, heads for an expanding, tails for a collapsing universe. The discovery of the redshift ruled out the contracting universe, so cosmologists turned their attention toward expanding models.

All sorts of expanding models were constructed. The Russian Alexander Friedmann and the Belgian Abbé Georges Lemaître were responsible for two early models. Some of these models assume a closed space (positive curvature), some an open space (negative curvature), some leave open the question of whether space is open or closed. Eddington devised a model and wrote a very readable book about it, *The Expanding Universe.* His model is essentially the same as Einstein's, closed like the surface of a vast four-dimensional balloon, but expanding uniformly in all three of its spatial dimensions. Today astronomers doubt that space closes on itself. The density of matter in space seems to be insufficient to account for such an overall positive curvature. Astronomers prefer the open or infinite universe of overall negative curvature, like the surface of a saddle.

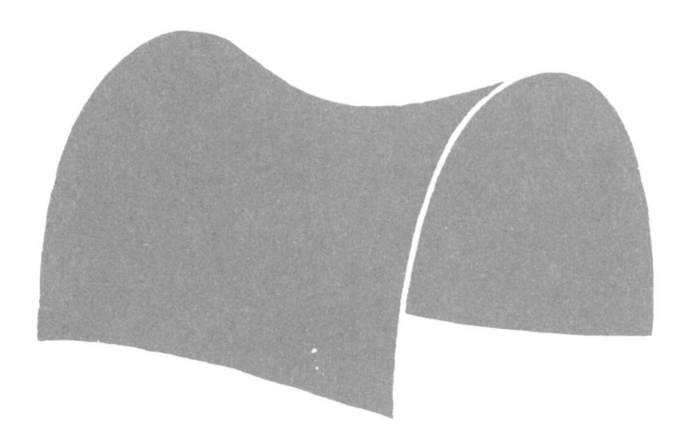

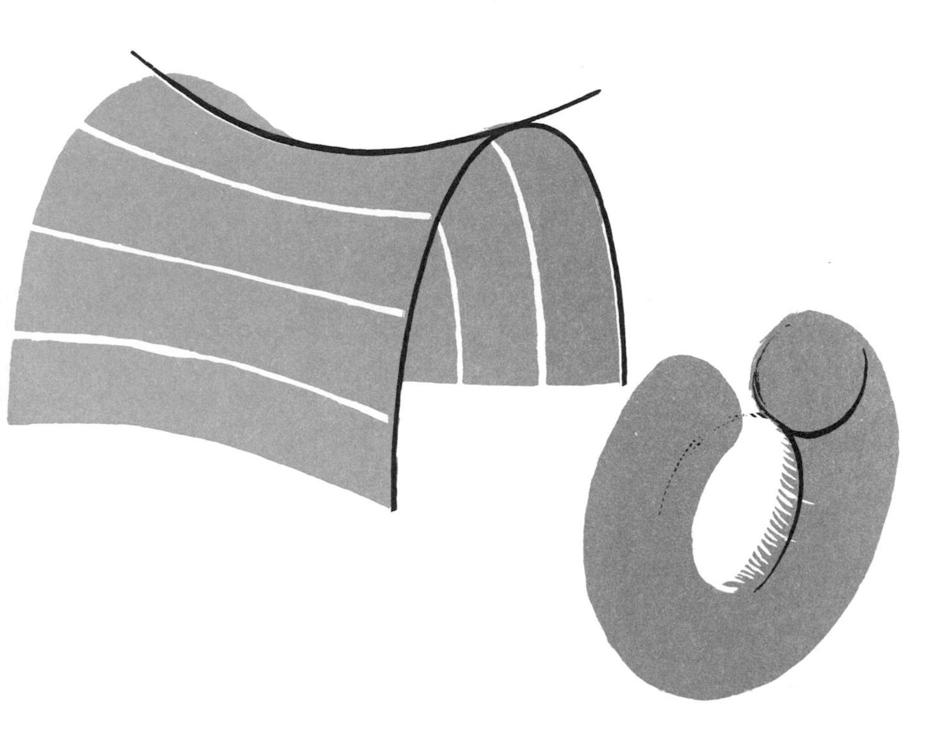

The reader must not suppose that because the surface of a sphere has positive curvature, the *inside* of a sphere's surface has negative curvature. The sphere's surface curvature is positive, whether viewed from one side or the other. The negative curvature of the saddle surface arises from the fact that at any point the surface is curving two different ways. It is concave if you run your hand over it from back to front, convex if you run your hand from side to side. One curvature is expressed by a positive number, the other by a negative number. The two numbers are multiplied to obtain the curvature of the surface at a given point. If at all points this number is negative, as it is sure to be if at all points the surface is curving two different ways, the surface is said to have negative curvature. The surface surrounding the hole of a torus (doughnut) is another familiar example of a surface with negative curvature. Such surfaces are, of course, only crude models of negatively curved three-dimensional space.

Perhaps more powerful telescopes will settle the question of whether the universe has positive, negative, or zero curvature. A telescope can see galaxies only within a certain spherical volume. If the galaxies are randomly distributed, and if space is Euclidian (zero curvature), the number of galaxies within such a sphere should always be proportional to the cube of the sphere's radius. In other words, if a telescope were built that could see twice as far into space as any previous telescope, the number of visible galaxies should jump from n to $8n$. If the jump were less than this, it would indicate a positive curving of the universe; if more than this, a negative curving.

One might think it would be the other way around, but consider the case of two-dimensional surfaces of positive and negative curvatures. Suppose a circle is cut from a flat sheet of rubber. On it are glued raisins at distances of a quarter-inch apart. To be formed into the surface of a sphere, the rubber must be *compressed* and many raisins brought closer together. In other words, if the raisins are to remain a quarter-inch apart on the spherical surface, *fewer* raisins will be needed. The reverse is true if the rubber is distorted into a saddle surface. This *stretches* the rubber and moves the raisins farther apart. To keep them a quarter-inch apart on the saddle surface, *more* raisins are needed. The moral of all this is, so runs a stale mathematical joke, that when you buy a bottle of beer, be sure to tell the clerk you want a bottle containing space that is curved negatively, not positively!

The expanding-universe models made it unnecessary to retain Einstein's cosmological constant, the hypothetical force that keeps the stars from moving together. (Einstein later considered this concept of a cosmological constant the greatest mistake he ever made.) The new models also cleared up immediately the problem of Olbers' paradox about the brightness of the night sky. Einstein's static model had been of little help on this score. True, it has only a finite number of suns, but because of the closed character of its space, light from these suns is trapped into going round and round the universe forever, twisting this way and that as it is bent by local distortions of spacetime. This would light up the night sky as much as an infinity of suns unless one assumes that the cosmos is so young that light has been able to make only a limited number of round trips.

The notion of an expanding universe eliminates the paradox very simply. If the distant galaxies are moving away from the earth with a speed proportional to their distance, the effect is a dimming of the total amount of light reaching the earth. If a galaxy is far enough away, its speed will exceed that of light. Its light will never reach us at all. Many astronomers today seriously believe that if the universe were not expanding, there literally would be no difference between night and day.

The fact that distant galaxies may exceed the speed of light relative to the earth seems to violate the dictum that no material body can go faster than light. But as we saw in Chapter 4, this dictum holds only for conditions that meet the requirements of the special theory. In the general theory, it must be rephrased as the dictum that no signals can be transmitted faster than light. Still, there is considerable controversy over whether distant galaxies actually can pass through the light barrier, so to speak, and vanish forever from man's ability to see them even if he had the most powerful telescopes imaginable. Some experts think the speed of light *is* a limit here, so that the most distant galaxies would simply grow dimmer and dimmer without ever becoming totally invisible, provided man had sufficiently sensitive instruments for detecting them.

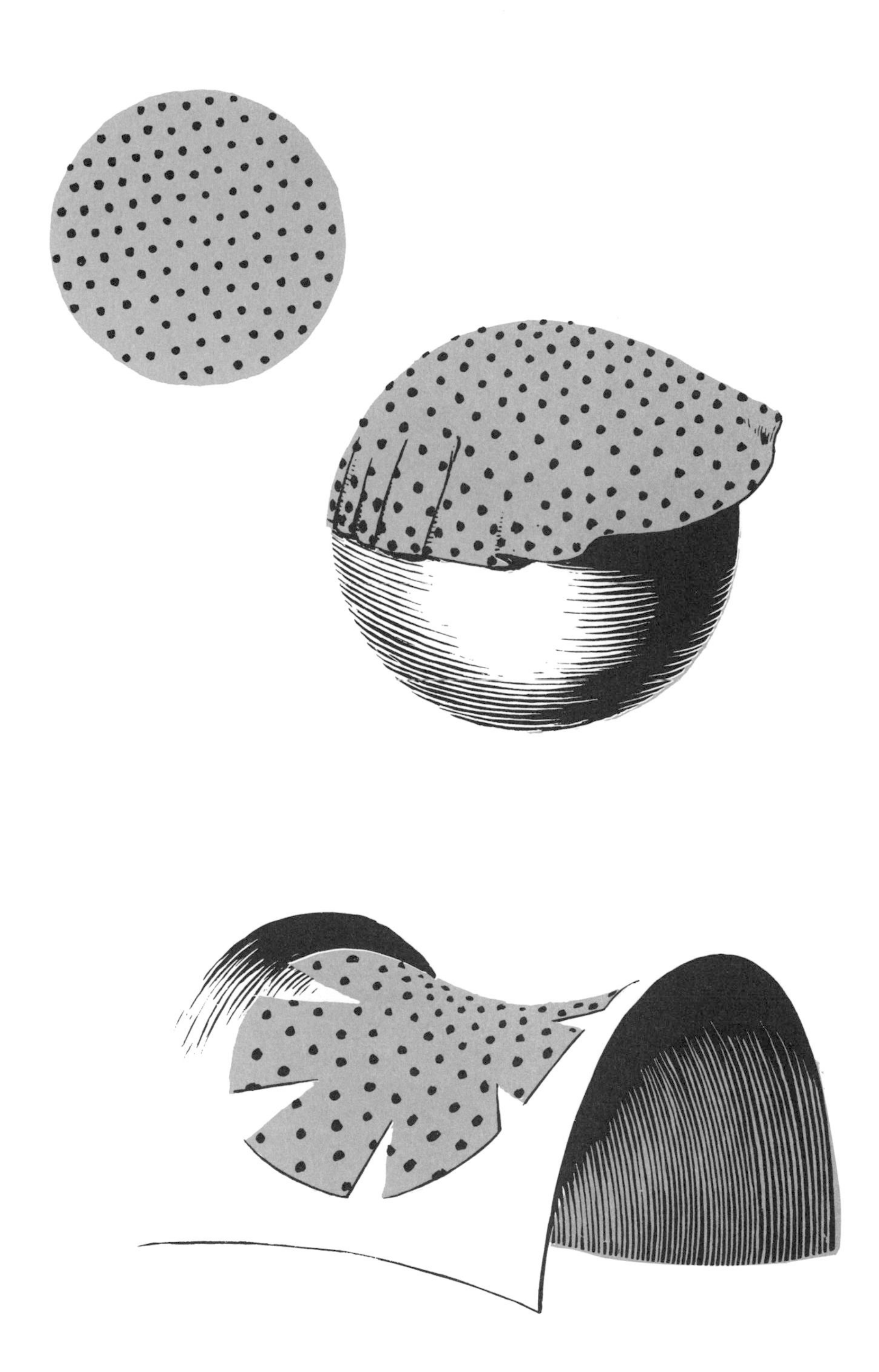

Old galaxies, someone was the first to say, never die. They just fade away. It is important to understand, however, that no galaxy actually vanishes in the sense that its matter disappears from the universe. It merely reaches a speed that makes it impossible, or almost impossible, for the earth's telescopes to detect it. The vanishing galaxy continues to be visible, of course, from all galaxies that surround it at closer range. For each galaxy there is an "optical horizon," a spherical boundary, beyond which its telescopes cannot penetrate. These spherical horizons are not the same for any two galaxies. Astronomers calculate that the point at which galaxies may be vanishing over *our* "rim" is about twice as far away as any present optical telescope can reach. If this assumption is correct, about one eighth of all the galaxies that will ever be visible are now being seen.

If the universe is expanding (regardless of whether it is flat or non-Euclidian, open or closed), then two tantalizing questions arise. What was the universe like if one goes back as far as possible in time? And what is going to happen to our universe if one goes forward as far as possible in time? These questions will be pondered in our final chapter. But first we must take a look at some spectacular new astronomical discoveries.

11

Quasars, Pulsars, and Black Holes

The past 35 years have seen a remarkable upsurge of interest in relativity theory. Scores of important new books and hundreds of technical papers on the subject have been published in all major languages. We saw how new laboratory techniques have made possible new confirmations of general relativity, but that is only one reason for the relativity explosion. The major reason is that in 1962, the very year that the first edition of this book was published, there began a series of incredible astronomical discoveries, all connected with relativity, that boggled the minds of astronomers and physicists.

The first of these extraordinary developments was the finding of a radically new kind of stellar object called the "quasar." The name stands for "quasi-stellar" radio source. The first quasar to be identified is known as 3C 273, the brightest and presumably the nearest of all the quasars. Australian radio astronomers had first pinpointed it as a source of strong radio waves, then Maarten Schmidt, at California's Palomar Observatory, identified the source with a fuzzy spot of light in the sky. An examination of the light's spectrum threw Schmidt into a state of shock. The redshift is so enormous that 3C 273 is apparently moving away from us with a velocity about 15 percent that of light, and so far away (more than a billion light-years) that there is no simple way to account for the strength of its radio emission. It is too big to be a star, too small and dense to be a galaxy, yet it is pouring out energy far greater than a galaxy.

Other quasars were soon detected, many of them even harder to comprehend than 3C 273. One seems to be moving away from us at a speed of more than 90 percent the speed of light, and throwing out energy about a hundred times that of a typical galaxy! Quasars have been found with the same stupendous redshifts but emitting little or no radio waves: "radio-quiet quasars" they are called. Hundreds of quasars have now been located and new ones are being discovered every week. A complete survey of the sky could turn up millions.

There is no consensus among experts on what quasars are, where they are, how they got there, or what is happening to them. At the moment the biggest debate is about *where* they are. Most cosmologists are persuaded that they are at the very rim of the seeable universe. This means that we are seeing objects about as far away as we can, and about as far backward in time. If this is true, the quasars must have come into being billions of years ago when the universe was just an infant.

A small number of astronomers violently disagree. They think the quasars are so near that they are actually within the cluster of galaxies to which our Milky Way belongs. If so, their enormous redshifts must have some unconventional explanation. Perhaps the quasars were expelled from within our galaxy when it was formed, and are still moving rapidly away from us. Perhaps their redshifts are produced by colossal gravitational forces, or by "tired light," or by some other cause not yet known. Any of these explanations, if confirmed, would throw modern cosmology into utter chaos.

Those who believe the quasars are nearby are no doubt responsible for the quip that short-sighted cosmologists think quasars are far away, whereas far-sighted cosmologists think they are close by. Astronomers like Halton C. Arp, James Terrell, and Geoffrey and Margaret Burbidge are leading the "far-sighted" group. They point to spots where two or more quasars seem to be associated with one another but have widely different redshifts. In one case a pair of quasars appear to be joined by a bridge of light, yet their redshifts are not the same. "Near-sighted" cosmologists argue that these are optical anomalies; that quasars which seem to be connected are actually remote from one another.

In 1971 astronomers found two radio sources, associated with a quasar, that seem to be separating from each other at more than nine times the speed of light! We saw in Chapter 4 how relativity permits an observer to chart the velocities of two objects, relative to one another, as close to twice the speed of light, but nine times the speed is an outright contradiction of relativity. On the other hand, if this quasar is near, estimates of the relative speeds of its two radio sources would drop to comprehensible values. Cosmologists who believe that this particular system is far away argue that the apparent speeds are an illusion caused by a "Christmas-tree effect." Objects, or parts of objects, that are not moving at all may be flashing on and off between observations, like Christmas-tree bulbs, to give an illusion of impossible motion.

The origin of quasar energy is also a dark mystery. The most popular theory is that the fantastic energy is produced by gravitational collapse, a process we will consider later in this chapter. George Gamow expressed the great mystification of astrophysicists by writing the following parody:

Twinkle, twinkle, quasi-star,
  Biggest puzzle from afar.
How unlike the other ones;
  Brighter than a billion suns.
Twinkle, twinkle, quasi-star,
  How I wonder what you are.

While astronomers were still wondering, and trying to recover their composure, they were rocked by an even stronger blow. This was the discovery of "pulsars." Pulsars are objects that send out radio pulses so accurately timed that when radio astronomers at the University of Cambridge first discovered them in 1967, they couldn't believe they were hearing anything natural in origin. For a few weeks they actually thought they had tuned in to some sort of message from intelligent life beyond the solar system.

One of the pulsars, NP 0531, is within the Crab Nebula in the constellation Taurus (the Bull). It is sending out beats of about thirty a second, with the precision of a clock that would be in error by only a second in many millions of years. When astronomers turned optical telescopes on the spot in the Crab Nebula where the pulses were coming from, they got another surprise. They found a point of light flashing on and off in perfect synchronization with the pulses! Of course, it had been flashing that way all along, but flashing so rapidly that it had appeared to the eye, and on photographs, as a steady point of light.

Since then more than a hundred pulsars have been spotted, some pulsing with visible light as well as radio waves. Their "tick" periods vary from 1/30 of a second (the tick of NP 0531) to almost four seconds. Using atomic clocks, astronomers have been able to measure these periods with accuracies of an eight-decimal fraction of a second, and to discover that all of them are slowing down by tiny amounts every year. Occasionally a pulsar undergoes a sudden increase in rate—astronomers call it a "glitch." NP 0531 has had several glitches since it was first identified. Like the quasars, new pulsars are constantly being found, and there may be millions in the sky.

Unlike quasars, pulsars are known to be small stellar objects inside our Milky Way galaxy. Most astronomers are convinced that they are rapidly spinning neutron stars, sending out radio beeps and sometimes flashing light in a manner similar to the rotating beam of a lighthouse. To explain what a neutron star is, let's take a quick look at the three primary ways a star in our galaxy can die. It turns out that the smaller a star, the gentler its death; the more massive, the more violent. Our sketch will deal only with the life histories of typical stars, but behind almost every sentence is a vast amount of theory that represents an extraordinary mixing of astronomy with relativity theory and particle physics.

First, let's consider what astronomers believe is the probable fate of a star about the size of our sun or smaller. Eventually such a star will burn up its hydrogen fuel and expand a few hundred times its former size to become what is called a red giant. The expansion will, of course, greatly thin the star's density. When our sun gets around to this, billions of years from now, it will probably swallow Mercury, Venus, and the earth. The reddish star Betelgeuse, the right "shoulder" of Orion, is a red giant.

Red giants produced by small stars remain in that condition only for a while. Eventually gravity causes them to contract to what is called a white dwarf, a star about the size of the earth but as massive as the sun. A piece of white dwarf the size of a pea would weigh (on earth) more than a hippopotamus. The enormous inpull of the white dwarf's gravity is balanced by the pressure of fast-moving electrons. The star's substance never loses its atomic structure. As time goes on, the white dwarf slowly cools to become a mass of cinders called a black dwarf.

Suppose that the original star is slightly larger than our sun but not twice as large. It, too, is likely to become a red giant. When it starts to contract, however, its greater mass causes it to pass a certain critical limit and the star may explode to become a supernova. When this happens, the explosion is visible in our sky as a new star much brighter than any others. The Crab Nebula is the remnant of just such an explosion. It occurred in 1054 and was so spectacular that it was carefully recorded by Oriental astronomers. Why it wasn't recorded in the Western world is still an historical mystery.

When a star explodes into a supernova, something very remarkable is believed to occur. J. Robert Oppenheimer and other physicists figured it all out on paper in 1938. The greater part of the star's mass shrinks in a few seconds to a star much smaller than the earth, a star of no more than ten to twenty kilometers wide. Gravitational forces are so intense within such a concentrated mass that the star becomes a million times as dense as the earth. A piece of the star the size of a marble weighs millions of tons. This is much too compact to allow preservation of atomic structure. Electrons and protons lose their identity and are squeezed into neutrons. The star becomes a neutron star.

Have you noticed that when a performing ice skater wants to spin like a top he (or she) starts whirling with arms outstretched, then suddenly the arms are pulled in close to the body? This sudden shift of mass to a smaller orbit causes the body to rotate faster. Exactly the same thing happens to a rapidly contracting neutron star. The spin it had as a red giant is enormously accelerated. The final result is a tiny neutron star, incredibly compact, and spinning faster than a ball on a juggler's fingertip. As it whirls it sends out radio pulses, sometimes accompanied by pulsing light. Exactly how the spinning star does this is still far from understood. The latest theories assume that the star's matter is in what is called a superfluid state—a state close to absolute zero temperature at which there is no viscosity or friction—and covered over by a thin, solid crust. Starquakes in this crust could be the cause of pulsar glitches.

Eventually the gas of the supernova fades. The explosion of the Crab Nebula was so recent, however, that its gaseous debris still glows in our sky. The little pulsar at the center is the neutron star that was once a sun slightly larger than our own.

Now we come to the third way a star can die, a way so bizarre that no one, before the theory of relativity was propounded, could have believed for a moment it was anything except the fantasy of a mad scientist or a writer of science fiction.

This third kind of fate awaits a star with an original mass much greater than the sun's, say, a mass of ten times or more. The huge star may go through its red-giant phase, but now, when it begins its gravitational collapse, its mass is so great and the forces of gravity so colossal that the pressure of electrons is insufficient to halt the star in a neutron stage. The implosion just keeps on going. It becomes a runaway, catastrophic collapse that transforms the star into what is called a black hole.

The most important thing to understand about a black hole is that it is smaller than its "Schwarzschild radius." As we learned back in Chapter 6, large masses alter the structure of spacetime so that rays of light passing close to the mass, taking geodesic paths, follow paths we see as curved in our ordinary three-dimensional space. The greater the mass, the greater this warping of spacetime and the greater the curvature of light. A few months after the theory of general relativity was published, Karl Schwarzschild, a German astronomer, proved that if gravitation compresses a mass within a certain radius (the distance dependent on the amount of mass), gravity would become so strong that no matter, radiation, or any sort of signal could escape from it. The radius of this sphere, into which anything can fall but from which nothing can escape, is the Schwarzschild radius.

Pierre Simon de Laplace, a French mathematician, pointed out as far back as 1798, using Newton's theory of gravity, that a star could be so massive with respect to its size that no light could escape from it. This was the first anticipation of what is now called a black hole. In 1939 Oppenheimer and his student Hartland S. Snyder made calculations similar to those of Laplace, but using the more refined formulas of relativity theory. They showed that if a star were sufficiently massive, it would indeed undergo a final catastrophic collapse to a density and size within the Schwarzschild radius. Since no light could escape from such a black hole, the star would become invisible. You couldn't see it by turning a big searchlight on it because the hole would simply absorb the light without reflecting any back. Either there are holes like this in the universe, someone has said, or there are holes in the theory of relativity.

For a star as massive as our sun, the Schwarzschild radius is between one and two kilometers. For an object like the earth, it is smaller than a marble.

A typical star, massive enough to collapse into a black hole, would produce a hole with a radius of only several kilometers. At the core of the hole is what mathematicians call a "singularity." Nothing is known about what happens to matter at such a spot because quantum mechanics no longer applies. Gravitational forces become infinite. Density and the curvature of spacetime become infinite. Particles of matter are literally crushed out of existence.

These ultimate distortions of space and time are reflected in the radically different ways that observers would view the final collapse. An observer safely outside a black hole would measure the time of the collapse as infinite, but to an observer on the star, his proper time would measure the collapse in milliseconds. An astronaut falling into a black hole would be instantly killed by the enormous tidal forces (see Chapter 5). They would compress him on all sides and pull him lengthwise into a fine filament that would approach zero thickness as he fell.

What would finally happen to his mass and energy? No one knows. Would it become pure spacetime? Would it become nothing? Is there a difference between spacetime and nothing? Perhaps a star that collapses into a black hole would go through what John Archibald Wheeler, the Princeton University physicist, calls a "worm hole" to re-enter another region of our universe. Or it might enter another universe lying outside our spacetime. Some physicists have speculated that our black holes are apertures through which energy is constantly pouring into another universe. The holes from which such energy emerges are called white holes. Is it possible that the black holes of some other cosmos are white holes at the centers of our quasars, holes through which energy is constantly pouring into *our* spacetime?

Another question, about which there is a frenzy of current debate, is whether a black hole can rotate. The assumption is that since stars can rotate, a star that collapsed into a black hole would produce a rotating hole. If so, the rotating hole might provide unlimited energy for a technologically advanced civilization. Wheeler has imagined such a society living on a gigantic shell that has been built around a rotating black hole the size of a mustard seed. He has even worked out a scheme by which garbage is dropped downward into the black abyss. The hole neatly disposes of the garbage, crushing it instantly out of existence, and politely sends back energy to supply all the needs of the civilization flourishing on the spherical shell.

Our universe, some physicists conjecture, may be dotted with millions of these "mini black holes." On June 30, 1908, there was a mysterious explosion in central Siberia. It was so cataclysmic that it blew over trees for a radius of more than thirty kilometers in all directions. To this day no one knows the cause of that monstrous blast. It was not a meteor, because no trace of a crater or buried meteorite has been found. Perhaps a comet struck the earth. Could it have been a mini black hole the size of a dust grain but weighing a billion tons? It might have struck the earth, then passed straight through the earth to emerge on the other side and continue through space.

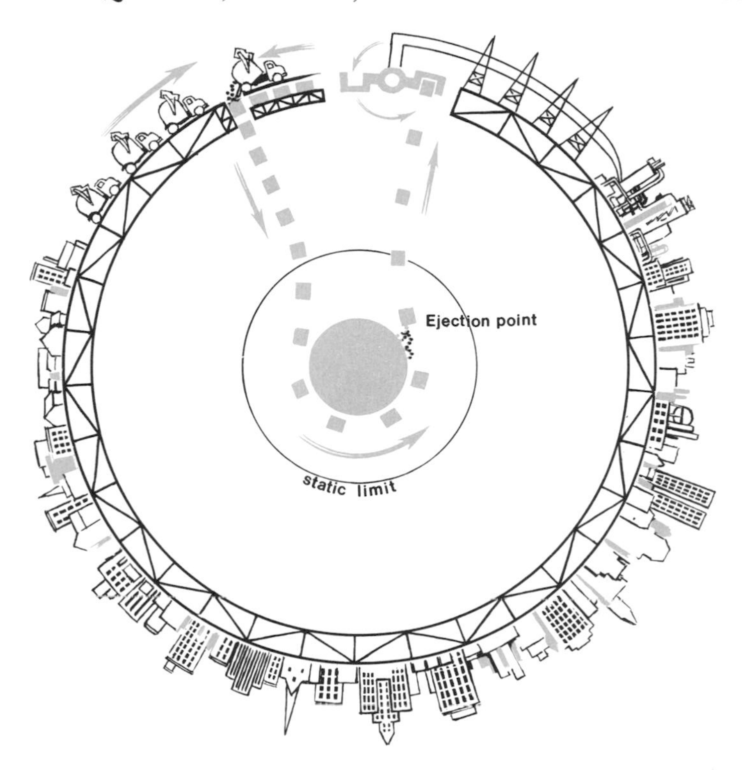

Is it possible for astronomers to detect the presence of a black hole in the sky? One way to look for a black hole is to look for an intense source of gravity waves. If a star or planet were to fall into a black hole, it would produce a burst of gravity waves. Some astronomers think that quasars are drawing their energy from black holes. Some think there may be black holes at the centers of galaxies. At the moment the most likely candidate for a black hole is the invisible X-ray source of Cygnus X-1 in the constellation Cygnus (the Swan). This is believed to be a system of two objects that revolve about once every five and a half days. The visible object is a supergiant star. The invisible component is apparently too massive to be a dwarf or a neutron star, and some astronomers have decided it must be a black hole. Skeptics are not so sure. Instead of a black hole there may be two objects, neither of them black holes.

Wheeler believes we are living inside a universe that will eventually stop expanding. It will then begin a contraction phase that eventually becomes runaway gravitational collapse. Finally the entire universe will vanish into its black singularity like the fabled Poof Bird that flies backward in ever decreasing circles until "Poof!"—it vanishes up its own posterior. Where it goes and what happens next are speculations that take us to our final chapter.

12

Beginning and End

Picture in your mind an expanding cosmos, then run the scene backward like a motion picture in reverse. It is apparent that there must have been a moment, in what Shakespeare once called the "dark backward and abysm of time," when an enormous amount of matter was concentrated in a very small space. Perhaps a great primeval explosion, many billions of years ago, started the whole process. This is the big-bang concept, which was first advocated by Lemaître (see Chapter 10), and which found its most able champion in George Gamow.

Gamow wrote a persuasive book, *The Creation of the Universe*, in defense of his theory. Lemaître thought the big bang took place about five billion years ago, but estimates of the age of the universe have also been expanding in recent years. It now appears that fifteen to twenty billion is a much better guess. At any rate, according to Gamow there was a time when all the matter in the universe was concentrated in one incredibly dense, uniform glob of concentrated matter which he liked to call Ylem (pronounced "eelem"; it is an old Greek word for primordial matter). How did the Ylem get there? Gamow thought it previously may have been spread out through the space of a *collapsing* universe. This period of the big squeeze is obviously a period about which information is hard to obtain. Like Lemaître's model, Gamow's model really begins with a bang. This is sometimes called the "moment of creation"—not in the sense of making something out of nothing, but in the sense of making a shape out of something previously shapeless. If belief in a creation out of nothing is preferred, *this* is as good a point as any, in Gamow's theory, to pick for it.

Just before the big bang, the temperature and pressure of the Ylem was incredibly high. Then came the monstrous, unimaginable explosion. Gamow's book will supply all the details of what may have happened after that. Eventually the stars congealed from the expanding dust and gas. The present expansion of the universe is the continuation of the motion imparted to matter by the initial explosion. Gamow believed that the motion will never stop.

In 1961, when I wrote the first edition of this book, the chief rival to the big-bang theory was the steady-state universe proposed in 1948 by three Cambridge University scientists: Hermann Bondi, Thomas Gold, and Fred Hoyle. The most persuasive defense of *this* theory is Hoyle's popular book *The Nature of the Universe*. Like Gamow's theory, the steady-state theory accepts the expansion of the universe and assumes space to be open and infinite rather than closed as in Eddington's model. Unlike Gamow's theory, it does not start with a bang. In fact, it does not start at all. Not by accident does the title of Hoyle's book differ from Gamow's only by the change of one word. Hoyle's cosmos has no moment of "creation"—or rather, it has, as we shall see, an infinity of small creations. As Hoyle expresses it elsewhere: "Every cluster of galaxies, every star, every atom had a beginning, but not the Universe itself. The Universe is something more than its parts, a perhaps unexpected conclusion."

The steady-state universe is always in the running, just the way it is now. Going back a hundred thousand billion years, the same types of evolving galaxies are found in any portion of the cosmos, containing the same types of aging stars, some with the same types of planets whirling around them, and on some of these planets, probably, similar types of life. There may be an infinity of planets on which at this very moment (whatever that may mean) intelligent creatures are sending their first astronauts into space. The cosmos is uniform, in an overall way, throughout an infinite space and an infinite time.

The universe is expanding, says the steady-state theory, but the expansion is not the aftermath of an explosion. It is due to an unknown repulsive force similar to Einstein's abandoned cosmological constant. Perhaps, some steady-staters argue, the force is caused by an infinitesimal difference between the proton's positive charge and the electron's negative charge. Atoms, hitherto considered neutral, would carry tiny charges, and since like charges repel, the universe would have a built-in tendency to expand. Whatever the nature of the repulsive force, it pushes the galaxies apart until finally they vanish over the "rim" as they pass through the light barrier. This disappearance is, of course, from the standpoint of an observer in our galaxy. When an observer on the earth sees Galaxy X and its neighbors fade away, observers in Galaxy X see *our* galaxy do the same thing.

An all-important question remains: If the universe has always been expanding and will keep on expanding forever, why doesn't it thin itself out? Clearly there is no way to maintain the steady state without assuming that new matter is constantly being created, perhaps in the form of hydrogen, the simplest of the elements. According to Hoyle (it is almost impossible to

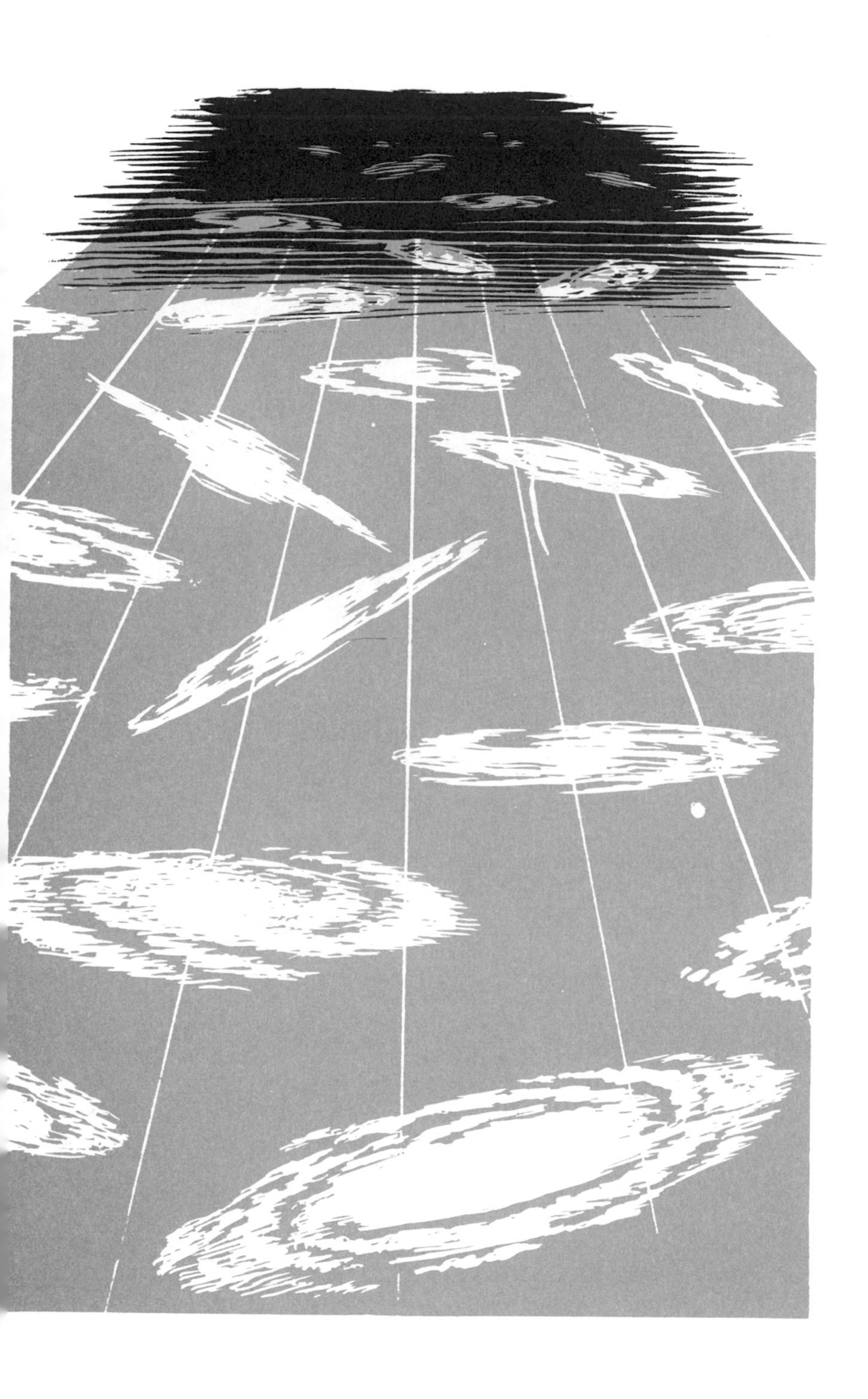

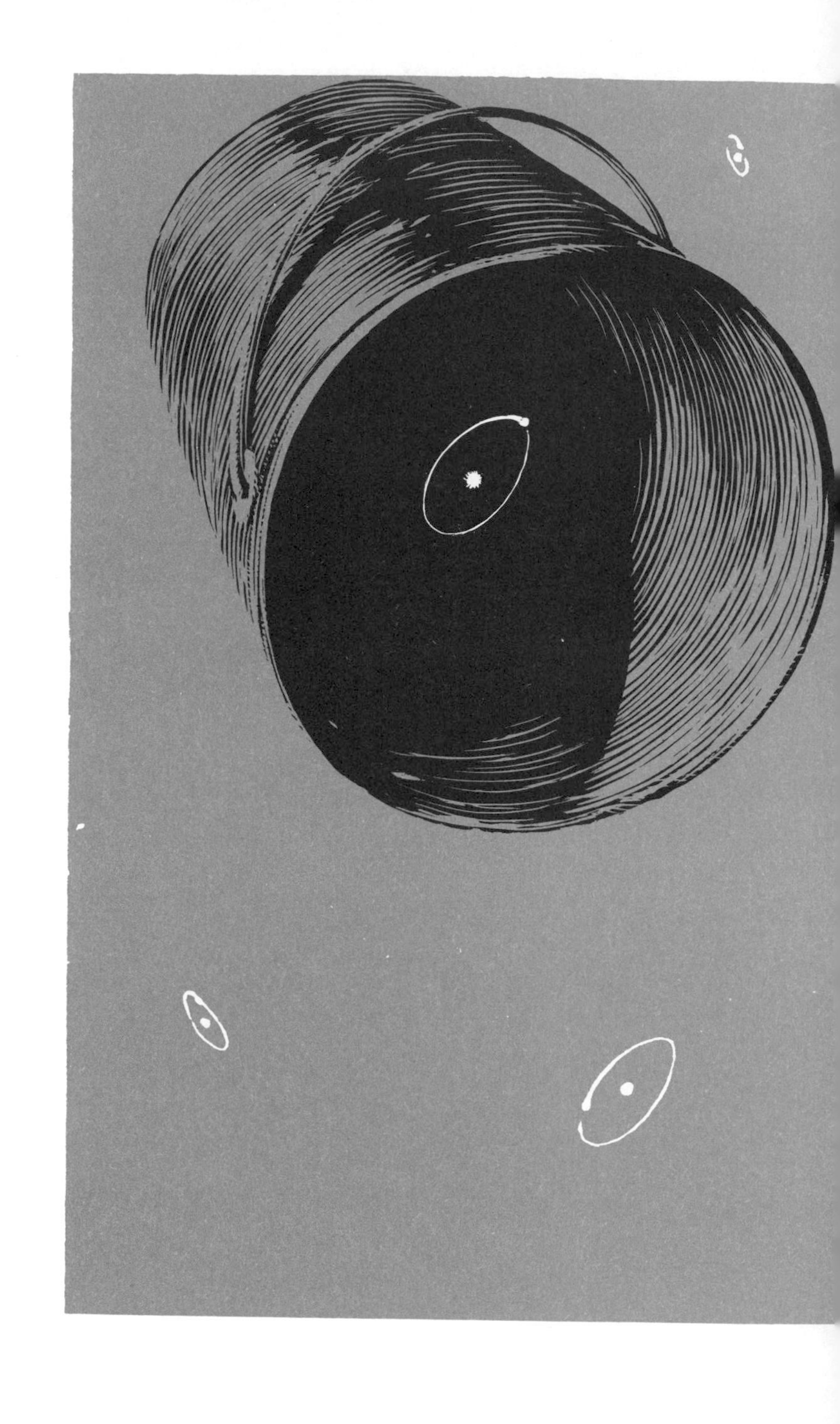

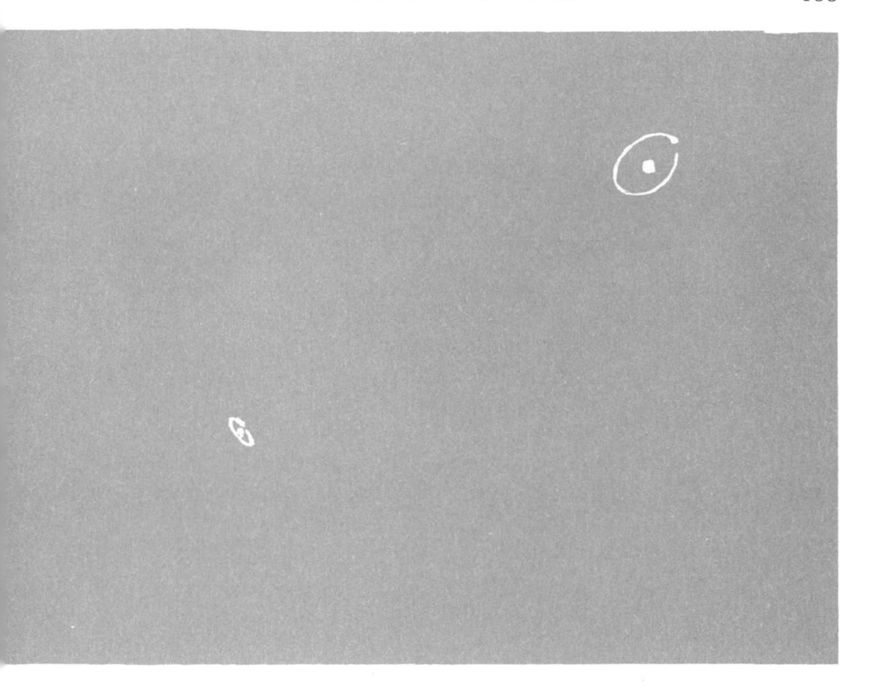

write about Hoyle's views without indulging in this obvious play on words), if one hydrogen atom per bucketful of space were to come into existence about every ten million years, it would keep the cosmos steady. Naturally, the rate at which matter forms has to be just such as to balance the thinning-out process.

Where do the hydrogen atoms come from? No one presumes to know. This is the point at which Hoyle's theory begins. If a belief in creation from nothing is maintained, *this* is the point in the steady-state theory where creation takes place, or rather, where it constantly takes place.

In 1961, the two rival theories, the big bang and the steady state, were running neck and neck. The values of the relevant parameters—variables needed for the construction of the models—simply were not known with enough precision to decide between the two theories. Relativity theory applied equally well to both, and both fitted equally well into the known facts about the universe—more accurately, what were *thought at the time* to be facts. But "facts" are hard to come by in cosmology, and estimates of the parameters are constantly altering. In the 1950s astronomers on each side of the debate wrote books and articles that made it appear as if all the evidence were on their side, and not much evidence on the side of their stubborn, out-of-date opponents.

Looking back on it now one can see that strong, sometimes unconscious emotional attitudes played a role. For many people there is something deeply disturbing about the notion of a universe coming into existence with a big explosion and then expanding forever until it freezes to death, and something cozy and comforting about a universe that is always the same. Hoyle and his followers were eloquent in expressing their emotional preference for this kind of universe.

For others it works the opposite way. Some people can think of nothing more ghastly than a universe forever expanding and forever remaining the same throughout an infinity of time and space. If there is such a thing as human immortality, G. K. Chesterton once wrote, perhaps it is part of God's mercy that he cuts it up for us into finite pieces so we can enjoy it. Maybe God himself, wanting to enjoy the spectacle of cosmic history, has to cut it up like a ribbon into finite pieces. At any rate, Gamow was equally frank about his emotional preference for the big bang. Is it possible that American culture, born in a recent revolution, inclines American astronomers toward a revolutionary beginning of the universe? Teller once suggested that the steady state was the dominant theory in England not only because it was the work of British cosmologists but because it expressed the British desire to maintain the status quo around the world.

Suddenly, in the mid-sixties, a funny thing happened to Hoyle on his way to the observatory. His steady-state theory faded like a galaxy vanishing from sight at the optical edge of the universe. The first big blow to his theory was the discovery of the quasars. Assuming that their redshifts do not have unconventional causes, they are structures existing only at the far edges of our cosmos. This means they were formed billions of years ago and haven't been forming since. There is no good way that the steady-state theory can account for them.

The greatest blow, the fatal blow, to the steady state was the discovery in 1965 that the universe is permeated with high-frequency waves on the borderline between microwave radio emissions and infrared light. It is sometimes called "black-body radiation" because black bodies at extremely low temperatures emit radio waves of this sort. The only good way astrophysicists can account for this radiation is to assume it is the residue of the great flash of light from the primordial fireball.

The infrared radiation was independently discovered by scientists at the Bell Laboratories in New Jersey and by physicists at Princeton University. It was Robert Dicke, the Princeton physicist, who had first proposed looking for such radiation, and his associates who found it were using a radiometer that Dicke had designed. It was a remarkable coincidence that the two groups of scientists, although working in laboratories near to each other, were unaware of each other's work, yet found the microwave radiation at about the same time.

The radiation "sea" is unquestionably there. It has a temperature of about 3 degrees on the absolute scale; a faint electromagnetic glow, a merest "whisper," left over from the big bang. At bang time the waves must have had short wavelengths, but they have been lengthened by the expansion of the universe over the past fifteen to twenty billion years.

The remarkable thing about this microwave radiation is its "isotropy"—that is, its uniformity in all spatial directions. It is this that rules out the possibility that the black-body radiation is coming from some single, unknown source. If such were the case, it could not be isotropic. The isotropy is so uniform that for the first time astronomers now have a way to measure the "absolute" motion of the earth. As we have seen, the earth moves around the sun, the sun moves through the Milky Way, the Milky Way rotates and moves within a cluster, and the cluster belongs to a moving supercluster of some 2,500 galaxies. Now that we know the universe is uniformly permeated with microwave radiation, we can use the Doppler effect to measure the earth's motion with respect to that radiation. We simply check on the amount of shift in various directions. Attempts to do just this are now under way, although no hard conclusions have yet been reached.

The reason I put "absolute" in quotes in the above paragraph is because it would be a measurement of the earth's motion with respect to the largest frame of reference we know—the universe itself. You must not suppose that this in any way violates relativity. One could just as legitimately assume the earth to be fixed and the entire universe, with its great spherical cloud of black-body radiation, to be moving. The equations are the same. Indeed, from the standpoint of relativity the choice of reference frame is arbitrary. Naturally, it is simpler to assume the universe is fixed and the earth moving than the other way around, but the two ways of talking about the earth's relative motion are two ways of saying the same thing.

It took a long while for proponents of the steady state to give up the theory. Dennis Sciama expressed himself movingly about it. "I must add that for me," he wrote in *Scientific American* (September 1967), "the loss of the steady-state theory has been a cause of great sadness. The steady-state theory has a sweep and beauty that for some unaccountable reason the architect of the universe appears to have overlooked. The universe in fact is a botched job, but I suppose we shall have to make the best of it."

The biggest botch, Sciama goes on to say, is the big bang itself. It was to avoid this "unpleasant singularity," he writes, that led to the steady-state theory. When Sciama was the guest at a *Scientific American* editorial luncheon, shortly before he wrote this, I heard him say that for years he had been trying to "weasel" out of the mounting evidence for the "unpleasant singularity" until finally he "ran out of weasels."

Hoyle seems not to have totally abandoned hope. For years he has proposed various weasels, some of them fantastic. Almost every year he puts forth a new scheme designed to restore some kind of steady state to everything, though not of course in the form of his dead theory. None of these new Hoylean speculations are taken seriously by the big-bangers. As George P. Thomson once said, criticizing Hoyle for his propensity to invent new laws to fit his theories, "There are many ways to solve chess problems if one is allowed to invent fresh moves for the pieces."

The fact that the big bang has won the day doesn't mean that cosmologists are now in substantial agreement. Far from it! It just means that the debate has shifted ground. Almost all experts now agree that our universe is expanding, that no new matter is entering it from higher spaces, and that it had its origin in a monstrous explosion some fifteen to twenty billion years ago.

The two biggest unanswered questions are about the beginning and the end. What happened before the explosion? What will eventually happen to the expanding universe?

All sorts of models can be constructed, but to decide between them we need more information about the fundamental parameters. The most important blank concerns the total amount of mass in the cosmos. Is there enough to halt the expansion and start the universe going the other way? The most recent attempts to calculate this mass give an amount that falls far short of what is necessary to halt the expansion, as well as far short of the amount necessary to "close" spacetime as in Einstein's original model or Eddington's modified version. Try as they will, astronomers are unable to find evidence for more than about 10 percent of the mass needed. If these estimates are correct, the open universe will go on expanding forever until it finally dissipates all its energy and expires from the cold.*

* See J. Richard Gott III, James E. Gunn, David N. Schramm and Beatrice M. Tinsley, "Will the Universe Expand Forever?," *Scientific American* (March 1976).

Cosmologists who cannot bear the thought that the universe will end in such a wintry whimper are forced to assume that somewhere in the universe there is enough "hidden mass" to close spacetime and eventually stop the universe from getting larger. Many theories have been proposed about the hiding places of this mass. The latest and most exotic is that it is hidden inside millions of mini black holes that were formed by the primordial explosion.

Let's assume the mass is there and that the universe will indeed go into a contracting phase. There is no way to halt such a contraction once it starts. The shrinking will finally become catastrophic and the universe will enter the singularity of a black hole. No one can think of any way to avoid the singularity, that dark abyss in which the density of the cosmos goes to infinity and matter is crushed out of existence.

Some cosmologists like to think the result will be another fireball. The new explosion will blow hydrogen into space as before, where it will condense into new galaxies, and the whole process will repeat itself. These are the "pulsating" or "oscillating" models. The idea was independently developed as a serious theory in 1919 by Howard P. Robertson and Richard C. Tolman. Of course the general notion, unsupported by scientific data, is much older. It underlies the concept of eternal reoccurrence that is part of many Eastern religions. Brahma, the Hindu creator god, inhales and exhales universes through his nose while Shiva, in a ring of fire, dances each old cosmos out of existence and dances in the new. Similar visions have been defended by a few Western philosophers. The ancient Stoics taught that the universe goes through endless cycles, each ending with everything dissolving in a central fireball. Nietzsche was obsessed by the same notion—the "ring of eternity," he called it—and defended it lyrically in *Thus Spake Zarathustra*.

Oscillating universes can be infinite in number, either closed structures within a spacetime of five dimensions, isolated from each other like bubbles in a foam, or parallel worlds in spacetimes of higher orders. This leads to a new kind of steady state.* Superspace is forever blowing bubbles. In recent years Hoyle himself has been attracted to this view, but it is Wheeler who has argued it with the greatest mathematical sophistication.

* If we go back to cosmologies that predate the discovery of the expansion of the universe, there are many "steady-state" models, starting with Aristotle's. W. D. Macmillan, the Chicago astronomer who opposed relativity (see Chapter 9), had a nonrelativistic steady-state theory in which the redshift is caused by light evaporating into space and re-emerging as hydrogen atoms. In its modern sense, "steady state" is limited to relativistic theories that take cosmic expansion into account.

In Wheeler's vision, our spacetime can be described by a point in an unthinkably vast superspace. Occasionally a portion of superspace ties itself in a complicated knot and there is an explosion which creates a universe of three spatial dimensions. Random factors enter the explosion, so that each time a new universe is born it has its own peculiar set of constants, particles, and laws. An infinity of different universes are forever coming into existence for a while, expanding, then contracting back to oblivion. We are living in one that happened to explode in just such a way as to produce particles and laws which permitted the universe to evolve complicated structures (us) capable of contemplating itself.

A curious historical footnote to all this is that Edgar Allan Poe had just such a vision. He described it in his last published work, a small but remarkable book called *Eureka* that came out in 1848. At the time, Poe's cosmology seemed bizarre to his contemporaries. Today it reads as if it had been written by one of Wheeler's students!

A universe begins, said Poe, when God creates a "primordial particle" out of nothing. From it, matter is "irradiated" spherically in all directions, in the form of an "inexpressibly great yet limited number of unimaginably yet not infinitely minute atoms." As the universe expands, gravity slowly gains the upper hand, and the matter condenses (as in Laplace's nebular hypothesis, which Poe admired) to form stars and planets. This theory, wrote Poe (in a phrase often applied to relativity), is too beautiful not to be true.

Our universe, Poe argued (restating Olbers' paradox), is finite, otherwise the entire sky would be blazing with starlight. However, it is only one of an infinity of universes, the others so "unspeakably distant" that no light passes between them. These cosmic bubbles are forever isolated from one another, and there is no way that an intelligence in one universe can ever become aware of another. Each cosmos has its own deity.

Eventually gravity halts the expansion and the cosmos starts to contract. Finally all matter returns to its original Unity; that is, it becomes nothing again. The final "globe of globes will instantaneously disappear," and the God of that universe will remain "all in all." The deity will then start another creation with (and it is here that Poe comes so remarkably close to Wheeler) "a new and perhaps totally different series of conditions." This cyclic process goes on "for ever, and for ever, and for ever; a novel Universe swelling into existence, and then subsiding into nothingness at every throb of the Heart Divine."

Poe's pulsating universe is now the favored model of many cosmologists. There are, of course, other models, some advanced seriously, some in jest. Hoyle once said that he had invented dozens of models so bizarre that he never published them, even though each was consistent with present estimates of the relevant parameters. There are models in which space twists back on itself like a Moebius strip (a one-sided surface formed by giving a strip of paper a half twist and joining the ends). If you travel once around such a universe, you find yourself back where you started, only everything is reversed as in a mirror. Of course you can go around once more and straighten everything out. The mathematician Kurt Gödel published in 1949 a strange nonexpanding model in which every point in space is rotating the same way around an axis. All the axes are parallel, and to an observer at any spot, the entire universe seems to be rotating around him in the same direction.

The "kinematic relativity" model of the Oxford University astronomer Edward A. Milne is perhaps the most bizarre of all. It introduces two essentially different kinds of time. In terms of one time, the universe is infinite in age and size, not expanding at all. In terms of the other time, it is finite in size and has been expanding only since a moment of creation. It is a matter of convenience which kind of time is taken as basic.

The English mathematician Edmund Whittaker* once proposed (as a joke) a diminishing-universe theory in which our finite cosmos is now contracting and its matter continually vanishing back into wherever it comes from in Hoyle's theory. The world eventually fades completely away. "This would have the recommendation," Whittaker writes, "of supplying a very simple picture of the final destiny of the universe." Of course, such a theory would have to explain why we see a galactic redshift instead of a violet one, but this is not hard to account for. All we have to do is borrow one of de Sitter's devices and assume that time is getting faster. (As a friend, Sidney Margulies, has pointed out, this might explain why, as one grows older, the years seem to slip by like months. They *are* slipping by like months!) Light that reaches the earth from a distant galaxy would then be light from the galaxy as it appeared millions of years ago when its light vibrated more slowly. This could produce a redshift large enough to more than balance the Doppler shift to the violet. Of course, the farther the galaxy, the older and redder it would appear.

The fact that a diminishing model can be constructed shows how flexible the equations of relativity are. They can be fitted to scores of different cosmic models, all of which account fairly well for everything that can at present be observed. It is interesting to find the English philosopher Francis Bacon writing in his *Novum Organum* in 1620: "Many hypotheses with regard to the heavens can be formed, differing in themselves, and yet sufficiently according with the phenomena." Modern cosmology is un-

* Sir Edmund Whittaker wrote a two-volume *History of the Theories of Aether and Electricity* (1900–1926). It is a monumental, valuable account, but marred by a curious attempt to minimize Einstein's contributions. The theory of relativity is regarded throughout as the creation of Lorentz. For an attempt to explain Whittaker's blindness, see "Poincaré, Einstein, and the Theory of Special Relativity," by Jeremy J. Gray, in *The Mathematical Intelligencer* (Vol. 7, 1995, pp. 65ff).

changed in this respect, though the amount of phenomena observed is much greater; therefore, there are grounds for supposing that modern models are closer to the truth than the old. Of course the fashionable cosmic models of a hundred years from now, based on astronomical data not known at the moment, may be wildly unlike any model now taken seriously.

It is with this humbling thought in mind that so many writers of popular books on modern cosmology, from Eddington to Sciama, have quoted from Book 8 of John Milton's *Paradise Lost.* The angel Raphael is speaking to Adam and Eve. Note how the relativity of the earth's motion is assumed.

> To ask or search I blame thee not; for Heaven
> Is as the Book of God before thee set,
> Wherein to read his wondrous works, and learn
> His seasons, hours, or days, or months, or years.
> This to attain, whether Heaven move or Earth
> Imports not, if thou reckon right; the rest
> From Man or Angel the great Architect
> Did wisely to conceal, and not divulge
> His secrets, to be scanned, by them who ought
> Rather admire. Or, if they list to try
> Conjecture, he his fabric of the Heavens
> Hath left to their disputes—perhaps to move
> His laughter at their quaint opinions wide
> Hereafter, when they come to model Heaven,
> And calculate the stars.

There is an amusing short tale by the Irish writer Lord Dunsany (in his book *The Man Who Ate the Phoenix*) in which Atlas explains to Dunsany what happened on the day when science made it no longer possible for mortals to believe in the old Greek model of the universe. Atlas admits that he had found his task rather dull and unpleasant. He was cold, because he had the earth's South Pole on the back of his neck, and his hands were always wet from the two oceans. But he remained at his task as long as people believed in him.

Then the world, Atlas says sadly, began to get "too scientific." He decided he was no longer needed. So he just put down the world and walked away.

"Yes," Atlas says, "Not without reflection, not without considerable reflection. But when I did it, I must say I was profoundly astonished; utterly astonished at what happened."

"And what did happen?"

"Simply nothing. Simply nothing at all."

In this book I have tried to tell the story of what happened on a more recent occasion when Newton's God of Absolute Motion, after a couple of prods by Einstein, put down the earth and walked away. Nothing much happened to the earth, at least not for a while. It continued to rotate on its axis, bulge at its equator, whirl around the sun. But something did happen to physics. Its power to explain, its power to predict, above all its power to alter the face of the earth for good or evil, became greater than it had ever been before.

Postscript, 1996

> "What do you know about Time?" she said.
> "Nothing," answered the old, black cat, though nobody spoke to him.
> —LORD DUNSANY, "The Avenger of Perdónis," in *Tales of Three Hemispheres*

Since the second edition of this book was published in 1976, hundreds of tests have been made, using advanced technology, of the special and general theories of relativity. Classic experiments such as the Michelson-Morley and the Kennedy-Thorndike have been tested with ever greater precision. Both theories have passed all tests with flying colors. For an excellent summary of these results I recommend physicist Clifford M. Will's excellent book *Was Einstein Right?* (Basic Books, revised edition 1993).

Will recalls in his book that in 1976 a famous astronomer at the California Institute of Technology advised a graduate student to avoid specializing in relativity because it "had so little connection with the rest of physics and astronomy." Kip Thorne, the student, wisely ignored this advice and is now a leading expert in the rapidly growing field of relativistic cosmology. Thanks to new telescopes and space probes, the universe has become a vast laboratory for testing Einstein's general theory. Relativity is inextricably bound up with observations of pulsars, quasars, and black holes, and with all the exciting new theories about the basic particles and the nature of the Big Bang that produced them.

In the light of observational and experimental results and the unification of gravity and inertia, the general theory of relativity is amazingly and beautifully simple. Professor Will recalls Einstein's joking remark that if tests ever decided against the theory it would only prove God made a mistake when he designed the universe. Of course Einstein knew that elegance is not enough to make a theory fertile. Early in the game he himself, as we have seen, proposed three ways of testing the basic ideas of general relativity. How much does light from distant stars bend when it passes close to the sun? Does the elliptical orbit of Mercury rotate on the plane at a rate which agrees with relativity? And is the wavelength of light shifted toward the red side of the spectrum when influenced by gravity?

Before 1960 all three tests had only weak confirmations. Repeated attempts to measure the bending of starlight, as it grazed the sun during a total eclipse, were marred by huge margins of error. Measurements did confirm bending, but the degree of bend was impossible to pin down. Even Newtonian physics, Will reminds us, predicts the bending of light by gravity, though at only half the amount required by relativity. Mercury's orbit seemed to support Einstein, but again other explanations could not be ruled out. The gravitational red shift of light had almost no empirical support.

In the 1960s, Will writes, physicists suddenly found themselves in possession of fantastically powerful new tools. Atomic clocks of various kinds made possible incredibly accurate measurements of time. Laser instruments were perfected. Larger radio and X-ray telescopes were built. Faster computers made it easier to analyze complex data. Radar and laser light could be bounced off mirrors on the moon and off planets and satellites. What Will calls a renaissance of interest in general relativity soon emerged. At first the solar system was the new testing "laboratory." In the 1970s the laboratory enlarged to regions far beyond our galaxy.

Professor Will makes an important distinction between the basic ideas of general relativity, which physicists now take for granted, and the ten tensor equations Einstein finally provided as a way of measuring the curvature of space-time. If by "general relativity" we mean those equations, then in the 1960s many rival theories, with slightly different equations, were proposed. The most important was a theory devised by Princeton's Robert Dicke and his former graduate student Carl Brans. The Brans-Dicke theory, as it was later called, accepted all the central ideas of general relativity but modified Einstein's field equations by adding a second field. As a consequence, it made predictions that differed slightly from Einstein's.

To repeat what I said earlier in Chapter 7, measurements of the sun's shape seemed to show that the sun was fatter at its equator than had been suspected, perhaps because its core rotated faster than its surface. When this oblateness was taken into account, the Brans-Dicke theory predicted the rotation of Mercury's orbit better than did Einstein's. In a chapter called "The Rise and Fall of the Brans-Dicke Theory" Will explains why knowledge of the sun's precise shape remains cloudy. The sun's brightness and the fact that it constantly throbs like a beating heart make its shape extremely difficult to determine. Some observations reported in 1985 seem to show that the sun's core rotates *more slowly* than its surface. In any case, support for the Brans-Dicke theory has been rapidly eroding.

The most precise measurements supporting Einstein over Brans-Dicke are described in the chapter "Do the Earth and the Moon Fall the Same?" Einstein's field equations require an absolute equivalence in the way all matter is influenced by gravity. "If we were to drop the Earth and a ball of aluminum in the gravitational field of some distant body," Will writes, "the two would fall at the same rate." A 1969 experiment, using lasers, verified that the earth and moon fall toward the sun with the same acceleration, and to a precision of one part in a hundred billion. Because the Brans-Dicke theory does not accept what is called the "strong equivalence principle," this test counted heavily against it. Had Einstein been told of its result, Will surmises, he would have replied, "Of course!"

Ephraim Fischbach of Purdue University announced in 1986 that he and his associates had found evidence for a hitherto undetected repulsive force which they call "hypercharge." If it exists, it would be much weaker than gravity—but could cause gravity to act differently on different kinds of matter. A feather would not fall in a vacuum with exactly the same acceleration as an iron ball. Such a new force would be a revolutionary challenge to the strong equivalence principle. As Will reports in a new chapter added to his second edition, "Is It Twilight Time for the Fifth Force?," sophisticated efforts to detect such a force have all been failures, although such attempts are continuing.

A test of relativity, not proposed by Einstein, involves the way gravity delays a light signal. Professor Will explains it with a rubber-sheet model. Put a heavy ball on the center of a flat elastic sheet supported at its perimeter. The ball will produce a depression—a three-dimensional distortion of the sheet's two-dimensional space. This causes a marble, placed anywhere outside the depression, to roll toward the ball. The ball does not pull the marble. The marble moves because of the sheet's curvature. If you imagine a light ray on the sheet, entering and leaving the depression, it will travel farther than it would if the sheet were flat. This is similar to what happens when light goes through a region strongly warped by a star's mass. Because the path has lengthened, there is what is called the Shapiro time delay, after Irwin Shapiro, who worked out the mathematics in the early 1960s. Complex measurements of this delay by Viking spacecraft have confirmed Einstein's field equations with an error of one part in one thousand. Will calls it "still the most accurate test of the theory ever performed."

In brief, Will's book answers the question posed by its title with a resounding yes. Einstein *was* right. Not only have his equations been confirmed over and over again, but the general theory has become indispensable for understanding the incredible new objects that modern telescopes have detected: the pulsars believed to be fast-spinning neutron stars and the far-distant quasars suspected of having black holes at their centers because there seems no other way to account for their enormous energy output. The day has long passed, writes Will, when cosmologists can remain ignorant of relativity. Every year astrophysicists find new phenomena that only the general theory can explain. The most recent are the powerful gravity fields outside our galaxy that act like mammoth lenses, magnifying and refracting what is seen through them. Such lenses were predicted by Einstein in 1936. In 1990 the Hubble Space Telescope recorded a beautifully defined pattern called the "Einstein Cross." It shows a galaxy whose gravitational field has acted as a lens to produce four images (above, below, and on each side) of a distant quasar.

Although tests of light bending as it passes close to the sun continue to be muddied by the sun's corona and other factors, extremely accurate tests confirming such bending have been made by radio telescopes (unknown of course to Einstein) of radio waves from stars and pulsars as the waves graze the sun.

New tests using radiation from pulsars while they orbit a star have strongly confirmed light's velocity to be independent of its source. The elliptical orbits of such binary pulsars, as they are called, swivel much faster than Mercury's orbit. Measurements of these precessions confirm Einstein's equations with amazing accuracy. Each time a pulsar goes behind its companion star, its radio signals are blocked. Measurements of the blocking show that the orbital periods are slowing down by just the amount predicted by relativity on the assumption that the pulsar is slowly losing energy by radiating gravity waves.

In spite of many recent attempts, gravity waves have yet to be detected, though few physicists doubt that they will be observed when the technology for seeing them improves. The quantized particle of the gravitational field, the graviton, also remains undetected. In modern superstring theory, all basic particles, which appear as mathematical points, are actually inconceivably tiny closed loops, like rubber bands, with enormous tensile strength. Their various modes of vibration, in higher space dimensions, produce the different particles. The theory remains controversial, but one of its greatest merits is that it explains the graviton as the simplest possible vibration of a superstring.

So far, this bizarre but elegant theory is the best candidate for unifying gravity and quantum mechanics although as yet there is no way to test the theory. (On superstrings, see my *New Ambidextrous Universe*, W. H. Freeman, 1990.)

The exact value of the "Hubble constant"—the rate at which the universe expands—is far from settled. It is not even known if it is a steady rate, or accelerating or decelerating. As Ronald Angel jingled in a letter to *Science News* (February 18, 1995):

> There was a stargazer named Hubble,
> Who said, "We expand like a bubble."
> But finding the rate,
> Was a source of debate,
> Dissension, contention, and trouble.

The value of Hubble's constant depends on the amount of hidden mass in the universe. Measurements of the rotation of spiral galaxies show that unless there is a hidden mass or "dark matter" within them, their speed of rotation would cause them to fly apart. Something is holding them together. For this and other reasons cosmologists are embarrassed to admit that about 90 percent of the universe's mass is as yet undetected! Perhaps the "hot" (fast-moving) neutrinos are not massless. Perhaps the universe contains large numbers of "machos"—invisible low-mass stars or giant planets like Jupiter. Undetected black holes are other candidates for the missing mass. Or there may be cold (slow-moving) dark matter made up of unseen exotic particles with such names as axions and wimps that have been conjectured but never found. The nature of the universe's hidden mass is the greatest mystery in today's cosmology.

Because no one knows how much mass pervades the universe, there is no way to know for sure whether the universe will expand forever (as most cosmologists believe) until it dies of the Big Chill, or whether there is enough mass to halt the expansion and start the universe going backward to its eventual demise in a Big Crunch. Models of oscillating universes, popular when this book's second edition was written, are now out of favor. The main reason is that no theorist has been able to come up with a good explanation of how a universe could get the energy to bounce back.

Although bouncing universes are no longer fashionable, there are plenty of wild speculations these days about a vast superspace in which many universes, perhaps an infinity, are popping in and out of existence like bubbles of foam on the river of time, as science-fiction writer Arthur C. Clarke once put it in a short story. Each new universe could have its own unique set of laws and constants, only a small number with laws that permit life to evolve. We are here, so goes an argument called the "anthropic principle," because our universe just happened to be one with laws that allowed you and me to exist.

I find it remarkable that on certain fundamental questions about the universe we are not much closer to answers than the ancients. Lucian, the Greek satirist of the second century, had these comments in an essay translated in the Oxford University Press four-volume edition of his works (Volume 3, page 130):

> How their [the philosophers'] theories conflict is soon apparent; next-door neighbours? no, they are miles apart. In the first place, their views of the world differ. Some say it had no beginning, and cannot end; others boldly talk of its creator and his procedure; what particularly entertained me was that these latter set up a contriver of the universe, but fail to mention where he came from, or what he stood on while about his elaborate task, though it is by no means obvious how there could be place or time before the universe came into being. . . . I wish I could give you their lucubrations on ideas and incorporeals, on finite and infinite. Over that point, now, there is fierce battle; some circumscribe the All, others will have it unlimited. At the same time they declare for a plurality of worlds, and speak scornfully of others who make only one.

The term "big bang" was coined by Fred Hoyle as one of derision. In *The Nature of the Universe* he called it an "old idea" that was "unsatisfactory even before detailed examination showed that it leads to serious difficulties. For when we look at our own galaxy there is not the smallest sign that such an explosion has occurred."

In the last chapter of the heavily revised 1960 edition of this book, Hoyle speculated on whether future observations will ever discredit his steady-state theory. "Is it likely that any astonishing new developments are lying in wait for us? Is it possible that cosmology of 500 years hence will extend as far beyond our present beliefs as our cosmology goes beyond that of Newton? It may surprise you to hear that I doubt whether this will be so." As someone has said, cosmologists are often wrong but seldom uncertain.

Sir Fred has never given up on his steady-state model, though his later versions of it have grown steadily more bizarre. In 1982 he and his associate and former student, the Indian physicist Jayant Narlikar, attacked the notion that the universe is expanding. The expansion is no more than an illusion, they argued, caused by the shrinking of atoms inside our measuring instruments! See "Was There Really a Big Bang?" by William Kaufman III, in *Science Digest* (March 1982).

Poincaré's thought experiment about doubling the universe, with which I opened this book, has since been the topic of much confusing debate. For the doubling to be meaningless it is necessary to include changes in time, in electrical charge, and in other factors that may or may not be related to size. The trouble is that the thought experiment is poorly defined. If, for example, an electron is a point, without internal structure, what does it mean to say it doubles in size? On the other hand, if the electron is a tiny loop as in superstring theory, doubling in size becomes meaningful. For an attack on Poincaré's doubling see "It Is False that Overnight Everything Has Doubled," by G. Schlesinger, in *Philosophical Studies* (Volume 15, 1964, pages 65–71). For a defense of the doubling, see Adolf Grünbaum's *Geometry and Chronometry in Philosophical Perspective* (University of Minnesota Press, 1986, pages 147–194).

One would have expected that since the death of Herbert Dingle, and in light of experimental confirmations of the twin paradox, there would no longer be any doubters of its validity. Surprisingly, this is not the case. Mendel Sachs, a physicist at the State University of New York at Buffalo, rejects the clock paradox, though on grounds entirely different from Dingle's. Dingle argued that relativity theory does indeed imply the paradox, but that this shows that relativity theory is false. Sachs accepts relativity, but thinks Einstein blundered in saying it entailed the paradox!

Sachs has called belief in the paradox "the scandal of 20th-century physics" and an example of antiscience comparable to astrology. In several papers during the 1970s he tried to show that the paradox did not follow from relativity theory, and in "Einstein's Later View of the Twin Paradox" (*Foundations of Physics*, Volume 15, 1985), he contends that Einstein himself later changed his mind about the twins' asymmetric aging. Waldyr A. Rodrigues, Jr., and Marcio A. F. Rosa, in "The Meaning of Time in the Theory of Relativity and Einstein's Later View of the Twin Paradox" (*Foundations of Physics*, Volume 19, 1989), give a proof of the so-called clock paradox, and also show that "Einstein never wrote a single line which endorses Sachs's misleading point of view."

In spite of the fact that no expert on relativity agrees with him, Sachs never gives up. In his recent book *Relativity in Our Times* (1993), Chapter 22 is titled "Relative Time and the Twin Paradox." Once again he denies that asymmetric aging occurs and once again insists that Einstein later changed his mind about the aging.

The twin paradox is closely related to the fact that a planet's elliptical orbit around a sun is a geodesic in the curved spacetime created by the sun's huge mass. As I said in Chapter 6, a planet orbits the sun along the shortest possible world line, but one with the longest elapsed proper time. If a moving object deviates from its geodesic it must follow a path that is not "straight" in spacetime, and therefore longer.

Imagine the earth to be propelled out of its orbit, swinging far out in space and back again to a spot just ahead of where it left its orbit, but moving so fast during its outward loop that its orbital period around the sun remains the same. To do this, the earth would for a while have to move very much faster than it does. This would retard its clocks. Although a clock on the sun would show the time of the earth's round trip to be the same as if it had not left its orbit, the proper time, measured by clocks on the earth, would show the trip around the sun to have taken a shorter time. The earth would thus correspond to the spaceship carrying the twin out and back. A person on the earth that deviated from its geodesic would be younger than he would have been had the earth remained on its normal path. This is what Bertrand

Russell meant when he said that objects like planets, moving along geodesics, display a "cosmic laziness." They follow the shortest path in spacetime, but take the longest proper time to move along it.

Einstein's cosmological constant, which he considered his greatest blunder, is now being seriously reconsidered, and may prove not to be a blunder after all! Observations in 1994 by Wendy Freedman of a type of star called a Cepheid variable suggest that the universe may be younger than previously supposed. Its age could be no more than 8 to 12 billion years, instead of 15 to 20. If true, this sharply contradicts strong evidence that some stars are at least 14 billion years old, making them older than the universe! How can one be older than one's mother?

A way out of this cosmological scandal would be to find, as Einstein had conjectured in 1917, a very weak propulsive force operating within matter. As we learned earlier, Einstein proposed this force to preserve his tidy, closed, steady-state cosmos from gravitational collapse. If such an antigravity force exists, it could mean that the universe started expanding very slowly, but after the galaxies and stars condensed it would cause a faster expansion that would make the universe appear younger than it really is. The situation is complex and bewildering. In any case there is no evidence yet for such a constant, or any good theory to explain its nature. Maybe in a few years the Hubble Telescope will resolve this curious crisis.

The most ambitious, most difficult project to test the general theory of relativity is known as Gravity Probe B, first proposed in 1959 by three Stanford physicists. Under the direction of Francis Everitt, the plan is to send four gyroscopes into space to orbit the earth. Each gyroscope is a quartz ball four centimeters in diameter, about the size of a table-tennis ball and so perfectly round that, as science writer James Trefil once put it, if it were enlarged to the size of the earth its highest mountain would be about a yard tall.

Each ball is coated with niobium. This allows each spinning ball, at a supercool temperature slightly above absolute zero, to remain suspended in a vacuum surrounded by superconducting electrical fields. Magnetometers will measure the degree to which the axes of the balls precess relative to the universe of stars. Because it would take almost a hundred thousand years for the axes to turn 360 degrees, you can imagine how precise the measurements must be to determine how much the axes drift in, say, one year of 5,000 orbits. The degree of shift will then either confirm Einstein's equations or disconfirm them and so perhaps support certain rival theories that make slightly different predictions.

NASA has already spent $140 million on Gravity Probe B. Another $50 million is needed in 1996, and more than $300 million may be required to complete the experiment. If Everitt can persuade Congress to provide this funding, and all else goes well, the little gyroscopes are expected to be launched in 1999. The worst outcome would be that deviations from Einstein's equations would be so slight as to be within experimental error.

The whole test would then have to be repeated!

Galileo and Newton made experiments, but the extraordinary thing about Einstein is that he made no experiments. Moreover, he was often unaware of significant tests that had strong bearings on his speculations. He just sat alone, thinking deeply about the secrets of the Old One, as he liked to call the universe. Newton was a devout Anglican who spent half his life struggling to unravel the mysteries of biblical prophecy. Einstein had no interest in any religion except in the sense that Spinoza, whose secular pantheism he admired, was religious. Yet he and Newton, in addition to their giant intellects and creative intuitions, shared a strong sense of wonder toward the Old One and of humility before the unanswerable riddle of existence. Both were Platonists in their conviction that what science knows is an infinitesimal portion of what it does not know.

Newton, in an often quoted passage, likened himself to a boy playing on the shore of a vast "ocean of truth," amusing himself by picking up here and there a smooth pebble or a patterned shell. Einstein made the same point with a different metaphor. He told an interviewer that he thought of himself as a child who has entered an enormous library, its books written in many languages. He takes down one volume and manages to translate a few pages. What a far cry from those now trying to persuade us that physics is on the brink of discovering Everything!

Index

This is primarily a name index. Topics are not included when they are so broad and all-pervasive that it would be meaningless to list dozens of page references. There are, for example, no entries for absolute, acceleration, energy, frame of reference, gravity, inertia, light, length, mass, motion, rest, relative, space, spacetime, thought experiment, time, and so on. For such topics the reader is urged to read the entire book.

MARTIN GARDNER is the prolific author of some sixty books about science, mathematics, philosophy, and literature. His science books include *Fads and Fallacies in the Name of Science* (Dover) and *The New Ambidextrous Universe* (Freeman). *The Night Is Large*, a collection of his essays written over the past fifty years, was published in 1996 by St. Martin's Press. The son of a geologist, Gardner was born in Tulsa in 1914 and is a graduate, majoring in philosophy, of the University of Chicago. He and his wife Charlotte live in the mountains of western North Carolina.

ANTHONY RAVIELLI is one of the nation's top graphic artists. Books he has illustrated include *Wonders of the Human Body, An Adventure in Geometry, My Ten Secrets of Bowling*, by Don Carter, and one other book by his friend Gardner, *Science Puzzlers*, a Dover paperback retitled *Entertaining Science Tricks with Everyday Objects*. For many years Ravielli illustrated covers and articles for *Sports Illustrated*. A native of Manhattan, he now lives in Stamford, Connecticut, with his wife, Georgia.

A CATALOG OF SELECTED DOVER BOOKS IN ALL FIELDS OF INTEREST

CATALOG OF DOVER BOOKS

ALICE'S ADVENTURES IN WONDERLAND, Lewis Carroll. Beloved classic about a little girl lost in a topsy-turvy land and her encounters with the White Rabbit, March Hare, Mad Hatter, Cheshire Cat, and other delightfully improbable characters. 42 illustrations by Sir John Tenniel. 96pp. 5³⁄₁₆ x 8¼. 0-486-27543-4

AMERICA'S LIGHTHOUSES: An Illustrated History, Francis Ross Holland. Profusely illustrated fact-filled survey of American lighthouses since 1716. Over 200 stations — East, Gulf, and West coasts, Great Lakes, Hawaii, Alaska, Puerto Rico, the Virgin Islands, and the Mississippi and St. Lawrence Rivers. 240pp. 8 x 10¾. 0-486-25576-X

AN ENCYCLOPEDIA OF THE VIOLIN, Alberto Bachmann. Translated by Frederick H. Martens. Introduction by Eugene Ysaye. First published in 1925, this renowned reference remains unsurpassed as a source of essential information, from construction and evolution to repertoire and technique. Includes a glossary and 73 illustrations. 496pp. 6⅛ x 9¼. 0-486-46618-3

ANIMALS: 1,419 Copyright-Free Illustrations of Mammals, Birds, Fish, Insects, etc., Selected by Jim Harter. Selected for its visual impact and ease of use, this outstanding collection of wood engravings presents over 1,000 species of animals in extremely lifelike poses. Includes mammals, birds, reptiles, amphibians, fish, insects, and other invertebrates. 284pp. 9 x 12. 0-486-23766-4

THE ANNALS, Tacitus. Translated by Alfred John Church and William Jackson Brodribb. This vital chronicle of Imperial Rome, written by the era's great historian, spans A.D. 14-68 and paints incisive psychological portraits of major figures, from Tiberius to Nero. 416pp. 5³⁄₁₆ x 8¼. 0-486-45236-0

ANTIGONE, Sophocles. Filled with passionate speeches and sensitive probing of moral and philosophical issues, this powerful and often-performed Greek drama reveals the grim fate that befalls the children of Oedipus. Footnotes. 64pp. 5³⁄₁₆ x 8 ¼. 0-486-27804-2

ART DECO DECORATIVE PATTERNS IN FULL COLOR, Christian Stoll. Reprinted from a rare 1910 portfolio, 160 sensuous and exotic images depict a breathtaking array of florals, geometrics, and abstracts — all elegant in their stark simplicity. 64pp. 8⅜ x 11. 0-486-44862-2

THE ARTHUR RACKHAM TREASURY: 86 Full-Color Illustrations, Arthur Rackham. Selected and Edited by Jeff A. Menges. A stunning treasury of 86 full-page plates span the famed English artist's career, from *Rip Van Winkle* (1905) to masterworks such as *Undine, A Midsummer Night's Dream,* and *Wind in the Willows* (1939). 96pp. 8⅜ x 11. 0-486-44685-9

THE AUTHENTIC GILBERT & SULLIVAN SONGBOOK, W. S. Gilbert and A. S. Sullivan. The most comprehensive collection available, this songbook includes selections from every one of Gilbert and Sullivan's light operas. Ninety-two numbers are presented uncut and unedited, and in their original keys. 410pp. 9 x 12. 0-486-23482-7

THE AWAKENING, Kate Chopin. First published in 1899, this controversial novel of a New Orleans wife's search for love outside a stifling marriage shocked readers. Today, it remains a first-rate narrative with superb characterization. New introductory Note. 128pp. 5³⁄₁₆ x 8¼. 0-486-27786-0

BASIC DRAWING, Louis Priscilla. Beginning with perspective, this commonsense manual progresses to the figure in movement, light and shade, anatomy, drapery, composition, trees and landscape, and outdoor sketching. Black-and-white illustrations throughout. 128pp. 8⅜ x 11. 0-486-45815-6

THE BATTLES THAT CHANGED HISTORY, Fletcher Pratt. Historian profiles 16 crucial conflicts, ancient to modern, that changed the course of Western civilization. Gripping accounts of battles led by Alexander the Great, Joan of Arc, Ulysses S. Grant, other commanders. 27 maps. 352pp. 5⅜ x 8½. 0-486-41129-X

BEETHOVEN'S LETTERS, Ludwig van Beethoven. Edited by Dr. A. C. Kalischer. Features 457 letters to fellow musicians, friends, greats, patrons, and literary men. Reveals musical thoughts, quirks of personality, insights, and daily events. Includes 15 plates. 410pp. 5⅜ x 8½. 0-486-22769-3

BERNICE BOBS HER HAIR AND OTHER STORIES, F. Scott Fitzgerald. This brilliant anthology includes 6 of Fitzgerald's most popular stories: "The Diamond as Big as the Ritz," the title tale, "The Offshore Pirate," "The Ice Palace," "The Jelly Bean," and "May Day." 176pp. 5⅜ x 8½. 0-486-47049-0

BESLER'S BOOK OF FLOWERS AND PLANTS: 73 Full-Color Plates from Hortus Eystettensis, 1613, Basilius Besler. Here is a selection of magnificent plates from the *Hortus Eystettensis,* which vividly illustrated and identified the plants, flowers, and trees that thrived in the legendary German garden at Eichstätt. 80pp. 8⅜ x 11. 0-486-46005-3

THE BOOK OF KELLS, Edited by Blanche Cirker. Painstakingly reproduced from a rare facsimile edition, this volume contains full-page decorations, portraits, illustrations, plus a sampling of textual leaves with exquisite calligraphy and ornamentation. 32 full-color illustrations. 32pp. 9⅜ x 12¼. 0-486-24345-1

THE BOOK OF THE CROSSBOW: With an Additional Section on Catapults and Other Siege Engines, Ralph Payne-Gallwey. Fascinating study traces history and use of crossbow as military and sporting weapon, from Middle Ages to modern times. Also covers related weapons: balistas, catapults, Turkish bows, more. Over 240 illustrations. 400pp. 7¼ x 10⅛. 0-486-28720-3

THE BUNGALOW BOOK: Floor Plans and Photos of 112 Houses, 1910, Henry L. Wilson. Here are 112 of the most popular and economic blueprints of the early 20th century — plus an illustration or photograph of each completed house. A wonderful time capsule that still offers a wealth of valuable insights. 160pp. 8⅜ x 11. 0-486-45104-6

THE CALL OF THE WILD, Jack London. A classic novel of adventure, drawn from London's own experiences as a Klondike adventurer, relating the story of a heroic dog caught in the brutal life of the Alaska Gold Rush. Note. 64pp. 5$\frac{3}{16}$ x 8¼. 0-486-26472-6

CANDIDE, Voltaire. Edited by Francois-Marie Arouet. One of the world's great satires since its first publication in 1759. Witty, caustic skewering of romance, science, philosophy, religion, government — nearly all human ideals and institutions. 112pp. 5$\frac{3}{16}$ x 8¼. 0-486-26689-3

CELEBRATED IN THEIR TIME: Photographic Portraits from the George Grantham Bain Collection, Edited by Amy Pastan. With an Introduction by Michael Carlebach. Remarkable portrait gallery features 112 rare images of Albert Einstein, Charlie Chaplin, the Wright Brothers, Henry Ford, and other luminaries from the worlds of politics, art, entertainment, and industry. 128pp. 8⅜ x 11. 0-486-46754-6

CHARIOTS FOR APOLLO: The NASA History of Manned Lunar Spacecraft to 1969, Courtney G. Brooks, James M. Grimwood, and Loyd S. Swenson, Jr. This illustrated history by a trio of experts is the definitive reference on the Apollo spacecraft and lunar modules. It traces the vehicles' design, development, and operation in space. More than 100 photographs and illustrations. 576pp. 6¾ x 9¼. 0-486-46756-2

A CHRISTMAS CAROL, Charles Dickens. This engrossing tale relates Ebenezer Scrooge's ghostly journeys through Christmases past, present, and future and his ultimate transformation from a harsh and grasping old miser to a charitable and compassionate human being. 80pp. 5³⁄₁₆ x 8¼. 0-486-26865-9

COMMON SENSE, Thomas Paine. First published in January of 1776, this highly influential landmark document clearly and persuasively argued for American separation from Great Britain and paved the way for the Declaration of Independence. 64pp. 5³⁄₁₆ x 8¼. 0-486-29602-4

THE COMPLETE SHORT STORIES OF OSCAR WILDE, Oscar Wilde. Complete texts of "The Happy Prince and Other Tales," "A House of Pomegranates," "Lord Arthur Savile's Crime and Other Stories," "Poems in Prose," and "The Portrait of Mr. W. H." 208pp. 5³⁄₁₆ x 8¼. 0-486-45216-6

COMPLETE SONNETS, William Shakespeare. Over 150 exquisite poems deal with love, friendship, the tyranny of time, beauty's evanescence, death, and other themes in language of remarkable power, precision, and beauty. Glossary of archaic terms. 80pp. 5³⁄₁₆ x 8¼. 0-486-26686-9

THE COUNT OF MONTE CRISTO: Abridged Edition, Alexandre Dumas. Falsely accused of treason, Edmond Dantès is imprisoned in the bleak Chateau d'If. After a hair-raising escape, he launches an elaborate plot to extract a bitter revenge against those who betrayed him. 448pp. 5³⁄₁₆ x 8¼. 0-486-45643-9

CRAFTSMAN BUNGALOWS: Designs from the Pacific Northwest, Yoho & Merritt. This reprint of a rare catalog, showcasing the charming simplicity and cozy style of Craftsman bungalows, is filled with photos of completed homes, plus floor plans and estimated costs. An indispensable resource for architects, historians, and illustrators. 112pp. 10 x 7. 0-486-46875-5

CRAFTSMAN BUNGALOWS: 59 Homes from "The Craftsman," Edited by Gustav Stickley. Best and most attractive designs from Arts and Crafts Movement publication — 1903–1916 — includes sketches, photographs of homes, floor plans, descriptive text. 128pp. 8¼ x 11. 0-486-25829-7

CRIME AND PUNISHMENT, Fyodor Dostoyevsky. Translated by Constance Garnett. Supreme masterpiece tells the story of Raskolnikov, a student tormented by his own thoughts after he murders an old woman. Overwhelmed by guilt and terror, he confesses and goes to prison. 480pp. 5³⁄₁₆ x 8¼. 0-486-41587-2

THE DECLARATION OF INDEPENDENCE AND OTHER GREAT DOCUMENTS OF AMERICAN HISTORY: 1775-1865, Edited by John Grafton. Thirteen compelling and influential documents: Henry's "Give Me Liberty or Give Me Death," Declaration of Independence, The Constitution, Washington's First Inaugural Address, The Monroe Doctrine, The Emancipation Proclamation, Gettysburg Address, more. 64pp. 5³⁄₁₆ x 8¼. 0-486-41124-9

THE DESERT AND THE SOWN: Travels in Palestine and Syria, Gertrude Bell. "The female Lawrence of Arabia," Gertrude Bell wrote captivating, perceptive accounts of her travels in the Middle East. This intriguing narrative, accompanied by 160 photos, traces her 1905 sojourn in Lebanon, Syria, and Palestine. 368pp. 5⅜ x 8½. 0-486-46876-3

A DOLL'S HOUSE, Henrik Ibsen. Ibsen's best-known play displays his genius for realistic prose drama. An expression of women's rights, the play climaxes when the central character, Nora, rejects a smothering marriage and life in "a doll's house." 80pp. 5³⁄₁₆ x 8¼. 0-486-27062-9

DOOMED SHIPS: Great Ocean Liner Disasters, William H. Miller, Jr. Nearly 200 photographs, many from private collections, highlight tales of some of the vessels whose pleasure cruises ended in catastrophe: the *Morro Castle, Normandie, Andrea Doria, Europa,* and many others. 128pp. 8⅞ x 11¾. 0-486-45366-9

THE DORÉ BIBLE ILLUSTRATIONS, Gustave Doré. Detailed plates from the Bible: the Creation scenes, Adam and Eve, horrifying visions of the Flood, the battle sequences with their monumental crowds, depictions of the life of Jesus, 241 plates in all. 241pp. 9 x 12. 0-486-23004-X

DRAWING DRAPERY FROM HEAD TO TOE, Cliff Young. Expert guidance on how to draw shirts, pants, skirts, gloves, hats, and coats on the human figure, including folds in relation to the body, pull and crush, action folds, creases, more. Over 200 drawings. 48pp. 8¼ x 11. 0-486-45591-2

DUBLINERS, James Joyce. A fine and accessible introduction to the work of one of the 20th century's most influential writers, this collection features 15 tales, including a masterpiece of the short-story genre, "The Dead." 160pp. 5³⁄₁₆ x 8¼. 0-486-26870-5

EASY-TO-MAKE POP-UPS, Joan Irvine. Illustrated by Barbara Reid. Dozens of wonderful ideas for three-dimensional paper fun — from holiday greeting cards with moving parts to a pop-up menagerie. Easy-to-follow, illustrated instructions for more than 30 projects. 299 black-and-white illustrations. 96pp. 8⅜ x 11. 0-486-44622-0

EASY-TO-MAKE STORYBOOK DOLLS: A "Novel" Approach to Cloth Dollmaking, Sherralyn St. Clair. Favorite fictional characters come alive in this unique beginner's dollmaking guide. Includes patterns for Pollyanna, Dorothy from *The Wonderful Wizard of Oz,* Mary of *The Secret Garden,* plus easy-to-follow instructions, 263 black-and-white illustrations, and an 8-page color insert. 112pp. 8¼ x 11. 0-486-47360-0

EINSTEIN'S ESSAYS IN SCIENCE, Albert Einstein. Speeches and essays in accessible, everyday language profile influential physicists such as Niels Bohr and Isaac Newton. They also explore areas of physics to which the author made major contributions. 128pp. 5 x 8. 0-486-47011-3

EL DORADO: Further Adventures of the Scarlet Pimpernel, Baroness Orczy. A popular sequel to *The Scarlet Pimpernel,* this suspenseful story recounts the Pimpernel's attempts to rescue the Dauphin from imprisonment during the French Revolution. An irresistible blend of intrigue, period detail, and vibrant characterizations. 352pp. 5³⁄₁₆ x 8¼. 0-486-44026-5

ELEGANT SMALL HOMES OF THE TWENTIES: 99 Designs from a Competition, Chicago Tribune. Nearly 100 designs for five- and six-room houses feature New England and Southern colonials, Normandy cottages, stately Italianate dwellings, and other fascinating snapshots of American domestic architecture of the 1920s. 112pp. 9 x 12. 0-486-46910-7

THE ELEMENTS OF STYLE: The Original Edition, William Strunk, Jr. This is the book that generations of writers have relied upon for timeless advice on grammar, diction, syntax, and other essentials. In concise terms, it identifies the principal requirements of proper style and common errors. 64pp. 5⅜ x 8½. 0-486-44798-7

THE ELUSIVE PIMPERNEL, Baroness Orczy. Robespierre's revolutionaries find their wicked schemes thwarted by the heroic Pimpernel — Sir Percival Blakeney. In this thrilling sequel, Chauvelin devises a plot to eliminate the Pimpernel and his wife. 272pp. 5³⁄₁₆ x 8¼. 0-486-45464-9

AN ENCYCLOPEDIA OF BATTLES: Accounts of Over 1,560 Battles from 1479 B.C. to the Present, David Eggenberger. Essential details of every major battle in recorded history from the first battle of Megiddo in 1479 B.C. to Grenada in 1984. List of battle maps. 99 illustrations. 544pp. 6½ x 9¼. 0-486-24913-1

ENCYCLOPEDIA OF EMBROIDERY STITCHES, INCLUDING CREWEL, Marion Nichols. Precise explanations and instructions, clearly illustrated, on how to work chain, back, cross, knotted, woven stitches, and many more — 178 in all, including Cable Outline, Whipped Satin, and Eyelet Buttonhole. Over 1400 illustrations. 219pp. 8⅜ x 11¼. 0-486-22929-7

ENTER JEEVES: 15 Early Stories, P. G. Wodehouse. Splendid collection contains first 8 stories featuring Bertie Wooster, the deliciously dim aristocrat and Jeeves, his brainy, imperturbable manservant. Also, the complete Reggie Pepper (Bertie's prototype) series. 288pp. 5⅜ x 8½. 0-486-29717-9

ERIC SLOANE'S AMERICA: Paintings in Oil, Michael Wigley. With a Foreword by Mimi Sloane. Eric Sloane's evocative oils of America's landscape and material culture shimmer with immense historical and nostalgic appeal. This original hardcover collection gathers nearly a hundred of his finest paintings, with subjects ranging from New England to the American Southwest. 128pp. 10⅝ x 9. 0-486-46525-X

ETHAN FROME, Edith Wharton. Classic story of wasted lives, set against a bleak New England background. Superbly delineated characters in a hauntingly grim tale of thwarted love. Considered by many to be Wharton's masterpiece. 96pp. 5³⁄₁₆ x 8 ¼. 0-486-26690-7

THE EVERLASTING MAN, G. K. Chesterton. Chesterton's view of Christianity — as a blend of philosophy and mythology, satisfying intellect and spirit — applies to his brilliant book, which appeals to readers' heads as well as their hearts. 288pp. 5⅜ x 8½. 0-486-46036-3

THE FIELD AND FOREST HANDY BOOK, Daniel Beard. Written by a co-founder of the Boy Scouts, this appealing guide offers illustrated instructions for building kites, birdhouses, boats, igloos, and other fun projects, plus numerous helpful tips for campers. 448pp. 5³⁄₁₆ x 8¼. 0-486-46191-2

FINDING YOUR WAY WITHOUT MAP OR COMPASS, Harold Gatty. Useful, instructive manual shows would-be explorers, hikers, bikers, scouts, sailors, and survivalists how to find their way outdoors by observing animals, weather patterns, shifting sands, and other elements of nature. 288pp. 5⅜ x 8½. 0-486-40613-X

FIRST FRENCH READER: A Beginner's Dual-Language Book, Edited and Translated by Stanley Appelbaum. This anthology introduces 50 legendary writers — Voltaire, Balzac, Baudelaire, Proust, more — through passages from *The Red and the Black, Les Misérables, Madame Bovary,* and other classics. Original French text plus English translation on facing pages. 240pp. 5⅜ x 8½. 0-486-46178-5

FIRST GERMAN READER: A Beginner's Dual-Language Book, Edited by Harry Steinhauer. Specially chosen for their power to evoke German life and culture, these short, simple readings include poems, stories, essays, and anecdotes by Goethe, Hesse, Heine, Schiller, and others. 224pp. 5⅜ x 8½. 0-486-46179-3

FIRST SPANISH READER: A Beginner's Dual-Language Book, Angel Flores. Delightful stories, other material based on works of Don Juan Manuel, Luis Taboada, Ricardo Palma, other noted writers. Complete faithful English translations on facing pages. Exercises. 176pp. 5⅜ x 8½. 0-486-25810-6

FIVE ACRES AND INDEPENDENCE, Maurice G. Kains. Great back-to-the-land classic explains basics of self-sufficient farming. The one book to get. 95 illustrations. 397pp. 5⅜ x 8½. 0-486-20974-1

FLAGG'S SMALL HOUSES: Their Economic Design and Construction, 1922, Ernest Flagg. Although most famous for his skyscrapers, Flagg was also a proponent of the well-designed single-family dwelling. His classic treatise features innovations that save space, materials, and cost. 526 illustrations. 160pp. 9⅜ x 12¼. 0-486-45197-6

FLATLAND: A Romance of Many Dimensions, Edwin A. Abbott. Classic of science (and mathematical) fiction — charmingly illustrated by the author — describes the adventures of A. Square, a resident of Flatland, in Spaceland (three dimensions), Lineland (one dimension), and Pointland (no dimensions). 96pp. 5³⁄₁₆ x 8¼. 0-486-27263-X

FRANKENSTEIN, Mary Shelley. The story of Victor Frankenstein's monstrous creation and the havoc it caused has enthralled generations of readers and inspired countless writers of horror and suspense. With the author's own 1831 introduction. 176pp. 5³⁄₁₆ x 8¼. 0-486-28211-2

THE GARGOYLE BOOK: 572 Examples from Gothic Architecture, Lester Burbank Bridaham. Dispelling the conventional wisdom that French Gothic architectural flourishes were born of despair or gloom, Bridaham reveals the whimsical nature of these creations and the ingenious artisans who made them. 572 illustrations. 224pp. 8⅜ x 11. 0-486-44754-5

THE GIFT OF THE MAGI AND OTHER SHORT STORIES, O. Henry. Sixteen captivating stories by one of America's most popular storytellers. Included are such classics as "The Gift of the Magi," "The Last Leaf," and "The Ransom of Red Chief." Publisher's Note. 96pp. 5³⁄₁₆ x 8¼. 0-486-27061-0

THE GOETHE TREASURY: Selected Prose and Poetry, Johann Wolfgang von Goethe. Edited, Selected, and with an Introduction by Thomas Mann. In addition to his lyric poetry, Goethe wrote travel sketches, autobiographical studies, essays, letters, and proverbs in rhyme and prose. This collection presents outstanding examples from each genre. 368pp. 5⅜ x 8½. 0-486-44780-4

GREAT EXPECTATIONS, Charles Dickens. Orphaned Pip is apprenticed to the dirty work of the forge but dreams of becoming a gentleman — and one day finds himself in possession of "great expectations." Dickens' finest novel. 400pp. 5³⁄₁₆ x 8¼. 0-486-41586-4

GREAT WRITERS ON THE ART OF FICTION: From Mark Twain to Joyce Carol Oates, Edited by James Daley. An indispensable source of advice and inspiration, this anthology features essays by Henry James, Kate Chopin, Willa Cather, Sinclair Lewis, Jack London, Raymond Chandler, Raymond Carver, Eudora Welty, and Kurt Vonnegut, Jr. 192pp. 5⅜ x 8½. 0-486-45128-3

HAMLET, William Shakespeare. The quintessential Shakespearean tragedy, whose highly charged confrontations and anguished soliloquies probe depths of human feeling rarely sounded in any art. Reprinted from an authoritative British edition complete with illuminating footnotes. 128pp. 5³⁄₁₆ x 8¼. 0-486-27278-8

THE HAUNTED HOUSE, Charles Dickens. A Yuletide gathering in an eerie country retreat provides the backdrop for Dickens and his friends — including Elizabeth Gaskell and Wilkie Collins — who take turns spinning supernatural yarns. 144pp. 5⅜ x 8½. 0-486-46309-5

HEART OF DARKNESS, Joseph Conrad. Dark allegory of a journey up the Congo River and the narrator's encounter with the mysterious Mr. Kurtz. Masterly blend of adventure, character study, psychological penetration. For many, Conrad's finest, most enigmatic story. 80pp. 5³⁄₁₆ x 8¼. 0-486-26464-5

HENSON AT THE NORTH POLE, Matthew A. Henson. This thrilling memoir by the heroic African-American who was Peary's companion through two decades of Arctic exploration recounts a tale of danger, courage, and determination. "Fascinating and exciting." — *Commonweal.* 128pp. 5⅜ x 8½. 0-486-45472-X

HISTORIC COSTUMES AND HOW TO MAKE THEM, Mary Fernald and E. Shenton. Practical, informative guidebook shows how to create everything from short tunics worn by Saxon men in the fifth century to a lady's bustle dress of the late 1800s. 81 illustrations. 176pp. 5⅜ x 8½. 0-486-44906-8

THE HOUND OF THE BASKERVILLES, Arthur Conan Doyle. A deadly curse in the form of a legendary ferocious beast continues to claim its victims from the Baskerville family until Holmes and Watson intervene. Often called the best detective story ever written. 128pp. 5³⁄₁₆ x 8¼. 0-486-28214-7

THE HOUSE BEHIND THE CEDARS, Charles W. Chesnutt. Originally published in 1900, this groundbreaking novel by a distinguished African-American author recounts the drama of a brother and sister who "pass for white" during the dangerous days of Reconstruction. 208pp. 5⅜ x 8½. 0-486-46144-0

THE HUMAN FIGURE IN MOTION, Eadweard Muybridge. The 4,789 photographs in this definitive selection show the human figure — models almost all undraped — engaged in over 160 different types of action: running, climbing stairs, etc. 390pp. 7⅞ x 10⅝. 0-486-20204-6

THE IMPORTANCE OF BEING EARNEST, Oscar Wilde. Wilde's witty and buoyant comedy of manners, filled with some of literature's most famous epigrams, reprinted from an authoritative British edition. Considered Wilde's most perfect work. 64pp. 5³⁄₁₆ x 8¼. 0-486-26478-5

THE INFERNO, Dante Alighieri. Translated and with notes by Henry Wadsworth Longfellow. The first stop on Dante's famous journey from Hell to Purgatory to Paradise, this 14th-century allegorical poem blends vivid and shocking imagery with graceful lyricism. Translated by the beloved 19th-century poet, Henry Wadsworth Longfellow. 256pp. 5³⁄₁₆ x 8¼. 0-486-44288-8

JANE EYRE, Charlotte Brontë. Written in 1847, *Jane Eyre* tells the tale of an orphan girl's progress from the custody of cruel relatives to an oppressive boarding school and its culmination in a troubled career as a governess. 448pp. 5³⁄₁₆ x 8¼. 0-486-42449-9

JAPANESE WOODBLOCK FLOWER PRINTS, Tanigami Kônan. Extraordinary collection of Japanese woodblock prints by a well-known artist features 120 plates in brilliant color. Realistic images from a rare edition include daffodils, tulips, and other familiar and unusual flowers. 128pp. 11 x 8¼. 0-486-46442-3

JEWELRY MAKING AND DESIGN, Augustus F. Rose and Antonio Cirino. Professional secrets of jewelry making are revealed in a thorough, practical guide. Over 200 illustrations. 306pp. 5⅜ x 8½. 0-486-21750-7

JULIUS CAESAR, William Shakespeare. Great tragedy based on Plutarch's account of the lives of Brutus, Julius Caesar and Mark Antony. Evil plotting, ringing oratory, high tragedy with Shakespeare's incomparable insight, dramatic power. Explanatory footnotes. 96pp. 5³⁄₁₆ x 8¼. 0-486-26876-4

THE JUNGLE, Upton Sinclair. 1906 bestseller shockingly reveals intolerable labor practices and working conditions in the Chicago stockyards as it tells the grim story of a Slavic family that emigrates to America full of optimism but soon faces despair. 320pp. 5$\frac{3}{16}$ x 8$\frac{1}{4}$. 0-486-41923-1

THE KINGDOM OF GOD IS WITHIN YOU, Leo Tolstoy. The soul-searching book that inspired Gandhi to embrace the concept of passive resistance, Tolstoy's 1894 polemic clearly outlines a radical, well-reasoned revision of traditional Christian thinking. 352pp. 5$\frac{3}{16}$ x 8$\frac{1}{4}$. 0-486-45138-0

THE LADY OR THE TIGER?: and Other Logic Puzzles, Raymond M. Smullyan. Created by a renowned puzzle master, these whimsically themed challenges involve paradoxes about probability, time, and change; metapuzzles; and self-referentiality. Nineteen chapters advance in difficulty from relatively simple to highly complex. 1982 edition. 240pp. 5$\frac{3}{8}$ x 8$\frac{1}{2}$. 0-486-47027-X

LEAVES OF GRASS: The Original 1855 Edition, Walt Whitman. Whitman's immortal collection includes some of the greatest poems of modern times, including his masterpiece, "Song of Myself." Shattering standard conventions, it stands as an unabashed celebration of body and nature. 128pp. 5$\frac{3}{16}$ x 8$\frac{1}{4}$. 0-486-45676-5

LES MISÉRABLES, Victor Hugo. Translated by Charles E. Wilbour. Abridged by James K. Robinson. A convict's heroic struggle for justice and redemption plays out against a fiery backdrop of the Napoleonic wars. This edition features the excellent original translation and a sensitive abridgment. 304pp. 6$\frac{1}{8}$ x 9$\frac{1}{4}$. 0-486-45789-3

LILITH: A Romance, George MacDonald. In this novel by the father of fantasy literature, a man travels through time to meet Adam and Eve and to explore humanity's fall from grace and ultimate redemption. 240pp. 5$\frac{3}{8}$ x 8$\frac{1}{2}$. 0-486-46818-6

THE LOST LANGUAGE OF SYMBOLISM, Harold Bayley. This remarkable book reveals the hidden meaning behind familiar images and words, from the origins of Santa Claus to the fleur-de-lys, drawing from mythology, folklore, religious texts, and fairy tales. 1,418 illustrations. 784pp. 5$\frac{3}{8}$ x 8$\frac{1}{2}$. 0-486-44787-1

MACBETH, William Shakespeare. A Scottish nobleman murders the king in order to succeed to the throne. Tortured by his conscience and fearful of discovery, he becomes tangled in a web of treachery and deceit that ultimately spells his doom. 96pp. 5$\frac{3}{16}$ x 8$\frac{1}{4}$. 0-486-27802-6

MAKING AUTHENTIC CRAFTSMAN FURNITURE: Instructions and Plans for 62 Projects, Gustav Stickley. Make authentic reproductions of handsome, functional, durable furniture: tables, chairs, wall cabinets, desks, a hall tree, and more. Construction plans with drawings, schematics, dimensions, and lumber specs reprinted from 1900s *The Craftsman* magazine. 128pp. 8$\frac{1}{8}$ x 11. 0-486-25000-8

MATHEMATICS FOR THE NONMATHEMATICIAN, Morris Kline. Erudite and entertaining overview follows development of mathematics from ancient Greeks to present. Topics include logic and mathematics, the fundamental concept, differential calculus, probability theory, much more. Exercises and problems. 641pp. 5$\frac{3}{8}$ x 8$\frac{1}{2}$. 0-486-24823-2

MEMOIRS OF AN ARABIAN PRINCESS FROM ZANZIBAR, Emily Ruete. This 19th-century autobiography offers a rare inside look at the society surrounding a sultan's palace. A real-life princess in exile recalls her vanished world of harems, slave trading, and court intrigues. 288pp. 5$\frac{3}{8}$ x 8$\frac{1}{2}$. 0-486-47121-7

ONE OF OURS, Willa Cather. The Pulitzer Prize–winning novel about a young Nebraskan looking for something to believe in. Alienated from his parents, rejected by his wife, he finds his destiny on the bloody battlefields of World War I. 352pp. 5³⁄₁₆ x 8¼. 0-486-45599-8

ORIGAMI YOU CAN USE: 27 Practical Projects, Rick Beech. Origami models can be more than decorative, and this unique volume shows how! The 27 practical projects include a CD case, frame, napkin ring, and dish. Easy instructions feature 400 two-color illustrations. 96pp. 8¼ x 11. 0-486-47057-1

OTHELLO, William Shakespeare. Towering tragedy tells the story of a Moorish general who earns the enmity of his ensign Iago when he passes him over for a promotion. Masterly portrait of an archvillain. Explanatory footnotes. 112pp. 5³⁄₁₆ x 8¼. 0-486-29097-2

PARADISE LOST, John Milton. Notes by John A. Himes. First published in 1667, *Paradise Lost* ranks among the greatest of English literature's epic poems. It's a sublime retelling of Adam and Eve's fall from grace and expulsion from Eden. Notes by John A. Himes. 480pp. 5³⁄₁₆ x 8¼. 0-486-44287-X

PASSING, Nella Larsen. Married to a successful physician and prominently ensconced in society, Irene Redfield leads a charmed existence — until a chance encounter with a childhood friend who has been "passing for white." 112pp. 5⅜ x 8½. 0-486-43713-2

PERSPECTIVE DRAWING FOR BEGINNERS, Len A. Doust. Doust carefully explains the roles of lines, boxes, and circles, and shows how visualizing shapes and forms can be used in accurate depictions of perspective. One of the most concise introductions available. 33 illustrations. 64pp. 5⅜ x 8½. 0-486-45149-6

PERSPECTIVE MADE EASY, Ernest R. Norling. Perspective is easy; yet, surprisingly few artists know the simple rules that make it so. Remedy that situation with this simple, step-by-step book, the first devoted entirely to the topic. 256 illustrations. 224pp. 5⅜ x 8½. 0-486-40473-0

THE PICTURE OF DORIAN GRAY, Oscar Wilde. Celebrated novel involves a handsome young Londoner who sinks into a life of depravity. His body retains perfect youth and vigor while his recent portrait reflects the ravages of his crime and sensuality. 176pp. 5³⁄₁₆ x 8¼. 0-486-27807-7

PRIDE AND PREJUDICE, Jane Austen. One of the most universally loved and admired English novels, an effervescent tale of rural romance transformed by Jane Austen's art into a witty, shrewdly observed satire of English country life. 272pp. 5³⁄₁₆ x 8¼. 0-486-28473-5

THE PRINCE, Niccolò Machiavelli. Classic, Renaissance-era guide to acquiring and maintaining political power. Today, nearly 500 years after it was written, this calculating prescription for autocratic rule continues to be much read and studied. 80pp. 5³⁄₁₆ x 8¼. 0-486-27274-5

QUICK SKETCHING, Carl Cheek. A perfect introduction to the technique of "quick sketching." Drawing upon an artist's immediate emotional responses, this is an extremely effective means of capturing the essential form and features of a subject. More than 100 black-and-white illustrations throughout. 48pp. 11 x 8¼. 0-486-46608-6

RANCH LIFE AND THE HUNTING TRAIL, Theodore Roosevelt. Illustrated by Frederic Remington. Beautifully illustrated by Remington, Roosevelt's celebration of the Old West recounts his adventures in the Dakota Badlands of the 1880s, from roundups to Indian encounters to hunting bighorn sheep. 208pp. 6¼ x 9¼. 0-486-47340-6

THE RED BADGE OF COURAGE, Stephen Crane. Amid the nightmarish chaos of a Civil War battle, a young soldier discovers courage, humility, and, perhaps, wisdom. Uncanny re-creation of actual combat. Enduring landmark of American fiction. 112pp. 5³⁄₁₆ x 8¼. 0-486-26465-3

RELATIVITY SIMPLY EXPLAINED, Martin Gardner. One of the subject's clearest, most entertaining introductions offers lucid explanations of special and general theories of relativity, gravity, and spacetime, models of the universe, and more. 100 illustrations. 224pp. 5⅜ x 8½. 0-486-29315-7

REMBRANDT DRAWINGS: 116 Masterpieces in Original Color, Rembrandt van Rijn. This deluxe hardcover edition features drawings from throughout the Dutch master's prolific career. Informative captions accompany these beautifully reproduced landscapes, biblical vignettes, figure studies, animal sketches, and portraits. 128pp. 8⅜ x 11. 0-486-46149-1

THE ROAD NOT TAKEN AND OTHER POEMS, Robert Frost. A treasury of Frost's most expressive verse. In addition to the title poem: "An Old Man's Winter Night," "In the Home Stretch," "Meeting and Passing," "Putting in the Seed," many more. All complete and unabridged. 64pp. 5³⁄₁₆ x 8¼. 0-486-27550-7

ROMEO AND JULIET, William Shakespeare. Tragic tale of star-crossed lovers, feuding families and timeless passion contains some of Shakespeare's most beautiful and lyrical love poetry. Complete, unabridged text with explanatory footnotes. 96pp. 5³⁄₁₆ x 8¼. 0-486-27557-4

SANDITON AND THE WATSONS: Austen's Unfinished Novels, Jane Austen. Two tantalizing incomplete stories revisit Austen's customary milieu of courtship and venture into new territory, amid guests at a seaside resort. Both are worth reading for pleasure and study. 112pp. 5⅜ x 8½. 0-486-45793-1

THE SCARLET LETTER, Nathaniel Hawthorne. With stark power and emotional depth, Hawthorne's masterpiece explores sin, guilt, and redemption in a story of adultery in the early days of the Massachusetts Colony. 192pp. 5³⁄₁₆ x 8¼. 0-486-28048-9

THE SEASONS OF AMERICA PAST, Eric Sloane. Seventy-five illustrations depict cider mills and presses, sleds, pumps, stump-pulling equipment, plows, and other elements of America's rural heritage. A section of old recipes and household hints adds additional color. 160pp. 8⅜ x 11. 0-486-44220-9

SELECTED CANTERBURY TALES, Geoffrey Chaucer. Delightful collection includes the General Prologue plus three of the most popular tales: "The Knight's Tale," "The Miller's Prologue and Tale," and "The Wife of Bath's Prologue and Tale." In modern English. 144pp. 5³⁄₁₆ x 8¼. 0-486-28241-4

SELECTED POEMS, Emily Dickinson. Over 100 best-known, best-loved poems by one of America's foremost poets, reprinted from authoritative early editions. No comparable edition at this price. Index of first lines. 64pp. 5³⁄₁₆ x 8¼. 0-486-26466-1

SIDDHARTHA, Hermann Hesse. Classic novel that has inspired generations of seekers. Blending Eastern mysticism and psychoanalysis, Hesse presents a strikingly original view of man and culture and the arduous process of self-discovery, reconciliation, harmony, and peace. 112pp. 5³⁄₁₆ x 8¼. 0-486-40653-9

SKETCHING OUTDOORS, Leonard Richmond. This guide offers beginners step-by-step demonstrations of how to depict clouds, trees, buildings, and other outdoor sights. Explanations of a variety of techniques include shading and constructional drawing. 48pp. 11 x 8¼. 0-486-46922-0

SMALL HOUSES OF THE FORTIES: With Illustrations and Floor Plans, Harold E. Group. 56 floor plans and elevations of houses that originally cost less than $15,000 to build. Recommended by financial institutions of the era, they range from Colonials to Cape Cods. 144pp. 8⅜ x 11. 0-486-45598-X

SOME CHINESE GHOSTS, Lafcadio Hearn. Rooted in ancient Chinese legends, these richly atmospheric supernatural tales are recounted by an expert in Oriental lore. Their originality, power, and literary charm will captivate readers of all ages. 96pp. 5⅜ x 8½. 0-486-46306-0

SONGS FOR THE OPEN ROAD: Poems of Travel and Adventure, Edited by The American Poetry & Literacy Project. More than 80 poems by 50 American and British masters celebrate real and metaphorical journeys. Poems by Whitman, Byron, Millay, Sandburg, Langston Hughes, Emily Dickinson, Robert Frost, Shelley, Tennyson, Yeats, many others. Note. 80pp. 5 3/16 x 8¼. 0-486-40646-6

SPOON RIVER ANTHOLOGY, Edgar Lee Masters. An American poetry classic, in which former citizens of a mythical midwestern town speak touchingly from the grave of the thwarted hopes and dreams of their lives. 144pp. 5 3/16 x 8¼. 0-486-27275-3

STAR LORE: Myths, Legends, and Facts, William Tyler Olcott. Captivating retellings of the origins and histories of ancient star groups include Pegasus, Ursa Major, Pleiades, signs of the zodiac, and other constellations. "Classic." — *Sky & Telescope.* 58 illustrations. 544pp. 5⅜ x 8½. 0-486-43581-4

THE STRANGE CASE OF DR. JEKYLL AND MR. HYDE, Robert Louis Stevenson. This intriguing novel, both fantasy thriller and moral allegory, depicts the struggle of two opposing personalities — one essentially good, the other evil — for the soul of one man. 64pp. 5 3/16 x 8¼. 0-486-26688-5

SURVIVAL HANDBOOK: The Official U.S. Army Guide, Department of the Army. This special edition of the Army field manual is geared toward civilians. An essential companion for campers and all lovers of the outdoors, it constitutes the most authoritative wilderness guide. 288pp. 5 3/16 x 8¼. 0-486-46184-X

A TALE OF TWO CITIES, Charles Dickens. Against the backdrop of the French Revolution, Dickens unfolds his masterpiece of drama, adventure, and romance about a man falsely accused of treason. Excitement and derring-do in the shadow of the guillotine. 304pp. 5 3/16 x 8¼. 0-486-40651-2

TEN PLAYS, Anton Chekhov. *The Sea Gull, Uncle Vanya, The Three Sisters, The Cherry Orchard,* and *Ivanov,* plus 5 one-act comedies: *The Anniversary, An Unwilling Martyr, The Wedding, The Bear,* and *The Proposal.* 336pp. 5 3/16 x 8¼. 0-486-46560-8

THE FLYING INN, G. K. Chesterton. Hilarious romp in which pub owner Humphrey Hump and friend take to the road in a donkey cart filled with rum and cheese, inveighing against Prohibition and other "oppressive forms of modernity." 320pp. 5⅜ x 8½. 0-486-41910-X

THIRTY YEARS THAT SHOOK PHYSICS: The Story of Quantum Theory, George Gamow. Lucid, accessible introduction to the influential theory of energy and matter features careful explanations of Dirac's anti-particles, Bohr's model of the atom, and much more. Numerous drawings. 1966 edition. 240pp. 5⅜ x 8½. 0-486-24895-X

TREASURE ISLAND, Robert Louis Stevenson. Classic adventure story of a perilous sea journey, a mutiny led by the infamous Long John Silver, and a lethal scramble for buried treasure — seen through the eyes of cabin boy Jim Hawkins. 160pp. 5 3/16 x 8¼. 0-486-27559-0